The Story of Eclipses

by George Chambers

CONTENTS.

THE STORY OF ECLIPSES.

CHAPTER I.

INTRODUCTION.

It may, I fear, be taken as a truism that "the man in the street" (collectively, the "general public") knows little and cares less for what is called physical science. Now and again when something remarkable happens, such as a great thunderstorm, or an earthquake, or a volcanic eruption, or a brilliant comet, or a total eclipse, something in fact which has become the talk of the town, our friend will condescend to give the matter the barest amount of attention, whilst he is filling his pipe or mixing a whisky and soda; but there is not in England that general attention given to the displays of nature and the philosophy of those displays, which certainly is a characteristic of the phlegmatic German. However, things are better than they used to be, and the forthcoming total eclipse of the Sun of May 28, 1900 (visible as it will be as a partial eclipse all over Great Britain and Ireland, and as a total eclipse in countries so near to Great Britain as Spain and Portugal, to say nothing of the United States), will probably not only attract a good deal of attention on the part of many millions of English-speaking people, but may also be expected to induce a numerically respectable remnant to give their minds and thoughts, with a certain amount of patient attention, to the Science and Philosophy of Eclipses.

There are other causes likely to co-operate in bringing this about. It is true that men's minds are more enlightened at the end of the 19th century than they were at the end of the 16th century, and that a trip to Spain will awaken vastly different thoughts in the year 1900 to those which would have been awakened, say in the year 1587; but for all that, a certain amount of superstition still lingers in the world, and total eclipses as well as comets still give rise to feelings of anxiety and alarm amongst ill-educated villagers even in so-called civilized countries. Some amusing illustrations of this will be presented in due course. For the moment let me content myself by stating the immediate aim of this little book, and the circumstances which have led to its being written. What those circumstances are will be understood generally from what has been said already. Its aim is the unambitious one of presenting in readable yet sound scientific language a popular account of eclipses of the Sun and Moon, and (very briefly) of certain kindred astronomical phenomena which depend upon causes in some degree similar to those which operate in connection with eclipses. These kindred phenomena are technically known as "Transits" and "Occultations." Putting these two matters entirely aside for the present, we will confine our attention in the first instance to eclipses; and as eclipses of the Sun do not stand quite on the same footing as eclipses of the Moon, we will, after stating the general circumstances of the case, put the eclipses of the Moon aside for a while.

CHAPTER II.

GENERAL IDEAS.

The primary meaning of the word "Eclipse" ([Greek: ekleipsis]) is a forsaking, quitting, or disappearance. Hence the covering over of something by something else, or the immersion of something in something; and these apparently crude definitions will be found on investigation to represent precisely the facts of the case.

Inasmuch as the Earth and the Moon are for our present purpose practically "solid bodies," each must cast a shadow into space as the result of being illuminated by the Sun, regarded as a source of light. What we shall eventually have to consider is: What results arise from the existence of these shadows according to the circumstances under which they are viewed? But before reaching this point, some other preliminary considerations must be dealt with.

The various bodies which together make up the Solar system, that is to say, in particular, those bodies called the "planets"--some of them "primary," others "secondary" (alias "Satellites" or "Moons")--are constantly in motion. Consequently, if we imagine a line to be drawn between any two at any given time, such a line will point in a different direction at another time, and so it may occasionally happen that three of these ever-moving bodies will come into one and the same straight line. Now the consequences of this state of things were admirably well pointed out nearly half a century ago by a popular writer, who in his day greatly aided the development of science amongst the masses. "When one of the extremes of the series of three bodies which thus assume a common direction is the Sun, the intermediate body deprives the other extreme body, either wholly or partially, of the illumination which it habitually receives. When one of the extremes is the Earth, the intermediate body intercepts, wholly or partially, the other extreme body from the view of the observers situate at places on the Earth which are in the common line of direction, and the intermediate body is seen to pass over the other extreme body as it enters upon or leaves the common line of direction. The phenomena resulting from such contingencies of position and direction are variously denominated Eclipses, Transits, and Occultations, according to the relative apparent magnitudes of the interposing and obscured bodies, and according to the circumstances which attend them."[1]

The Earth moves round the Sun once in every year; the Moon moves round the Earth once in every lunar month (27 days). I hope everybody understands those essential facts. Then we must note that the Earth moves round the Sun in a certain plane (it is nothing for our present purpose what that plane is). If the Moon as the Earth's companion moved round the Earth in the same plane, an eclipse of the Sun would happen regularly every month when the Moon was in "Conjunction" ("New Moon"), and also every month at the intermediate period there would be a total eclipse of the Moon on the occasion of every "Opposition" (or "Full Moon"). But inasmuch as the Moon's orbit does not lie in quite the same plane as the Earth's, but is inclined thereto at an angle which may be taken to average about 5-1/8? the actual facts are different; that is to say, instead of there being in every year about 25 eclipses (solar and lunar in nearly equal numbers), which there would be if the orbits had identical planes, there are only a very few eclipses in the year, never, under the most favourable circumstances, more than 7, and sometimes as few as 2. Nor are the numbers equally apportioned. In years where there are 7 eclipses, 5 of them may be of the Sun and 2 of the Moon; where there are only 2 eclipses, both must be of the Sun. Under no circumstances can there be in any one year more than 3 eclipses of the Moon, and in some years there will be none. The reasons for these diversities are of a technical character, and a full elucidation of them would not be of interest to the general reader. It may here be added, parenthetically, that the occasions will be very rare of there being 5 solar eclipses in one year. This last happened in 1823,[2] and will only happen once again in the next two centuries, namely in 1935. If a total eclipse of the Sun happens early in January there may be another in December of the same year, as in 1889 (Jan. 1 and Dec. 22). This will not happen again till 2057, when there will be total eclipses on Jan. 5 and Dec. 26. There is one very curious fact which may be here conveniently stated as a bare fact, reserving the explanation of it for a future page, namely, that eclipses of the Sun and Moon are linked together in a certain chain or sequence which takes rather more than 18 years to run out when the sequence recurs and recurs ad infinitum. In this 18-year period, which bears the name of the "Saros," there usually happen 70 eclipses, of which 41 are of the Sun and 29 of the Moon. Accordingly, eclipses of the Sun are more numerous than those of the Moon in the proportion of about 3 to 2, yet at any given place on the Earth more lunar eclipses are visible than solar eclipses, because the former when they occur are visible over the whole hemisphere of the Earth which is turned towards the Moon whilst the area over which a total eclipse of the Sun is visible is but a belt of the Earth no more than about 150 to 170 miles wide. Partial eclipses of the Sun, however, are visible over a very much wider area on either side of the path traversed by the Moon's shadow.

Confining our attention in the first instance to eclipses of the Sun, the diagrams fig. 2 and fig. 3 will make clear, with very little verbal description, the essential features of the two principal kinds of eclipses of the Sun. In these figures S represents the Sun, M the Moon and E the Earth. They are not, of course, even approximately drawn to scale either as to the size of the bodies or their relative distances, but this is a matter of no moment as regards the principles involved. M being in sunshine receives light on, as it were, the left hand side, which faces S the Sun. The shadow of the Moon cast into space is, in the particular case, thrown as regards its tip on to the Earth and is intercepted by the Earth. Persons at the moment situated on the Earth within the limits of this shadow will not see any part of the Sun at all; they will see, in fact, nothing but the Moon as a black disc with only such light behind and around it as may be reflected back on to the sky by the illuminated (but to the Earth invisible) hemisphere of the Moon, or as may proceed from the Sun's Corona (to be described presently). The condition of things therefore is that known as a "total" eclipse of the Sun so far as regards the inhabitants of the narrow strip of Earth primarily affected.

Fig. 3 represents nearly but not quite the same condition of things. Here the Earth and the Moon are in those parts of their respective orbits which put the two bodies at or near the maximum distance possible from the Sun and from one another. The Moon casts its usual shadow, but the tip does not actually reach any part of the Earth's surface. Or, in other words, to an observer on the Earth the Moon is not big enough to conceal the whole body of the Sun. The result is this; at the instant of central coincidence the Moon covers up only the centre of the Sun, leaving the outer edge all round uncovered.

This outer edge shows as a bright ring of light, and the eclipse is of the sort known as an "annular" eclipse of the Sun.[3] As the greatest breadth of the annulus can never exceed 1?minutes of arc, an annular eclipse may sometimes, in some part of its track, become almost or quite total, and vice vers?

The idea will naturally suggest itself, what exactly does happen to the inhabitants living outside (on the one side or the other) of the strip of the Earth where the central line of shadow falls? This depends in every case on circumstances, but it may be stated generally that the inhabitants outside the central line but within 1000 to 2000 miles on either side, will see a larger or smaller part of the Sun concealed by the Moon's solid body, simultaneously with the total concealment of the Sun to the favoured individuals who live, or who for the moment are located, within the limits of the central zone.

Now we must advance one stage in our conceptions of the movements of the Earth and the Moon, so far as regards the bearing of those movements on the question of eclipses. The Earth moves in a plane which is called the "Plane of the Ecliptic," and correspondingly, the Sun has an apparent annual motion in the same plane. The Moon moving in a different plane, inclined to the first mentioned one to the extent of rather more than 5? the Moon's orbit will evidently intersect the ecliptic in two places. These places of intersection are called "Nodes," and the line which may be imagined to join these Nodes is called the "Line of Nodes." When the Moon is crossing the ecliptic from the S. to the N. side thereof, the Moon is said to be passing through its "Ascending Node" ([Symbol: Ascending node]); the converse of this will be the Moon passing back again from the N. side of the ecliptic to the S. side, which is the "Descending Node" ([Symbol: Descending node]). Such changes of position, with the terms designating them, apply not only to the Moon in its movement round the Earth, but to all the planets and comets circulating round the Sun; and also to satellites circulating round certain of the planets, but with these matters we have no concern now.

FOOTNOTES:

[Footnote 1: D. Lardner, Handbook of Astronomy, 3rd ed., p. 288.]

[Footnote 2: But not one of them was visible at Greenwich.]

[Footnote 3: Latin Annulus, a ring.]

CHAPTER III.

THE "SAROS" AND THE PERIODICITY OF ECLIPSES.

To bring about an eclipse of the Sun, two things must combine: (1) the Moon must be at or near one of its Nodes; and (2), this must be at a time when the Moon is also in "Conjunction" with the Sun. Now the Moon is in Conjunction with the Sun (=?New Moon") 12 or 13 times in a year, but the Sun only passes through the Nodes of the Moon's orbit twice a year. Hence an eclipse of the Sun does not and cannot occur at every New Moon, but only occasionally. An exact coincidence of Earth, Moon, and Sun, in a straight line at a Node is not necessary to ensure an eclipse of the Sun. So long as the Moon is within about 18degrees of its Node, with a latitude of not more than 1degree34', an eclipse may take place. If, however, the distance is less than 15 及 and the latitude less than 1degree23' an eclipse must take place, though between these limits[4] the occurrence of an eclipse is uncertain and depends on what are called the "horizontal parallaxes" and the "apparent semi-diameters" of the two bodies at the moment of conjunction, in other words, on the nearness or "far-offness" of the bodies in question. Another complication is introduced into these matters by reason of the fact that the Nodes of the Moon's orbit do not occupy a fixed position, but have an annual retrograde motion of about 19 ☊, in virtue of which a complete revolution of the Nodes round the ecliptic is accomplished in 18 years 218-7/8 days (=?8.5997 years).

The backward movement of the Moon's Nodes combined with the apparent motion of the Sun in the ecliptic causes the Moon in its monthly course round the Earth to complete a revolution with respect to its Nodes in a less time (27.2 days) than it takes to get back to Conjunction with the Sun (29.5 days); and a curious consequence, as we shall see directly, flows from these facts and from one other fact. The other fact is to the Sun starting coincident with one of the Moon's Nodes, returns on the Ecliptic to the same Node in 346.6 days. The first named period of 27.2 days is called the "Nodical Revolution of the Moon" or "Draconic Month," the other period of 29.5 days is called the "Synodical Revolution of the Moon." Now the curious consequence of these figures being what they are is that 242 Draconic Months, 223 Lunations, and 19 Returns of the Sun to one and the same Node of the Moon's orbit, are all accomplished in the same time within 11 hours. Thus (ignoring refinements of decimals):--

Days Days. Years. Days. Hours.

242 times 27.2 = 6585.36 = 18 10 8?223 times 29.5 = 6585.32 = 18 10 7?19 times 346.6 = 6585.78 = 18 10 18?
The interpretation to be put upon these coincidences is this: that supposing Sun and Moon to start together from a Node they would, after the lapse of 6585 days and a fraction, be found again together very near the same Node. During the interval there would have been 223 New and Full Moons. The exact time required for 223 Lunations is such that in the case supposed the 223rd conjunction of the two bodies would happen a little before they reached the Node; their distance therefrom would be 28' of arc. And the final fact is that eclipses recur in almost, though not quite, the same regular order every 6585-1/3 days, or more exactly, 18 years, 10 days, 7 hours, 42 minutes.[5] This is the celebrated Chaldean "SAROS," and was used by the

ancients (and can still be used by the moderns in the way of a pastime) for the prediction of eclipses alike of the Sun and of the Moon.

* * * * *

At the end of a Saros period, starting from any date that may have been chosen, the Moon will be in the same position with respect to the Sun, nearly in the same part of the heavens, nearly in the same part of its orbit, and very nearly indeed at the same distance from its Node as at the date chosen for the terminus a quo of the Saros. But there are trifling discrepancies in the case (the difference of about 11 hours between 223 lunations and 19 returns of the Sun to the Moon's Node is one) and these have an appreciable effect in disturbing not so much the sequence of the eclipses in the next following Saros as their magnitude and visibility at given places on the Earth's surface. Hence, a more accurate succession will be obtained by combining 3 Saros periods, making 54 years, 31 days; while, best of all, to secure an almost perfect repetition of a series of eclipses will be a combination of 48 Saroses, making 865 years for the Moon; and of about 70 Saroses, or more than 1200 years for the Sun.

These considerations are leading us rather too far afield. Let us return to a more simple condition of things. The practical use of the Saros in its most elementary conception is somewhat on this wise. Given 18 or 19 old Almanacs ranging, say, from 1880 to 1898, how can we turn to account the information they afford us in order to obtain from them information respecting the eclipses which will happen between 1899 and 1917? Nothing easier. Add 18^y 10^d 7^h 42^m to the middle time of every eclipse which took place between 1880 and 1898 beginning, say, with the last of 1879 or the first of 1880, and we shall find what eclipses will happen in 1898 and 17 following years, as witness by way of example the following table:--

Error of Saros by d. h. m. Exact Calculation. MOON. 1879 Dec. 28 4 26 p.m. (Mag. 0.17) 18 10 7 42 --- (Mag. 0.16) 1898 Jan. 8 12 8 a.m. (civil time) +3 m.

d. h. m. SUN. 1880 Jan. 11 10 48 p.m. (Total) 18 10 7 42 --------------------------
------------------ (Total) 1898 Jan. 22 6 30 a.m. (civil time) -1 h. 7 m.

d. h. m. MOON. 1880 June 22 1 50 p.m. (Mag. Total) 18 11 7 42 ----------------
-------------------------- (Mag. 0.93) 1898 July 3 9 32 p.m. +35 m.

d. h. m. SUN. 1880 July 7 1 35 p.m. (Mag. Annular) 18 11 7 42 -------------------
----------------------- (Mag. Annular) 1898 July 18 9 17 p.m. +1 h. 10 m.

d. h. m. SUN. 1880 Dec. 2 3 11 a.m. (civil time). (Mag. 0.04) 18 11 7 42 --------
------------------------------------ (Mag. 0.02) 1898 Dec. 13 10 53 a.m. -1 h. 5 m.

d. h. m. MOON. 1880 Dec. 16 3 39 p.m. (Mag. Total) 18 11 7 42 ----------------
-------------------------- (Mag. Total) 1898 Dec. 27 11 21 p.m. -13 m.

d. h. m. SUN. 1880 Dec. 31 1 45 p.m. (Mag. 0.71) 18 11 7 42 ---------------------
---------------------- (Mag. 0.72) 1899 Jan. 11 9 27 p.m. -1 h. 11 m.

There having been 5 recurrences of Feb. 29 between Dec. 1879 and Jan. 1899, 5 leap years having intervened, we have to add an extra day to the Saros period in the later part of the above Table.[6]

Let us make another start and see what we can learn from the eclipses, say, of 1883.

Error of Saros by Exact Calculation. h. m. MOON 1883 April 22 11 39 a.m. (Mag. 0.8) 18 11 7 42 -- (Mag. Penumbral) 1901 May 3 7 21 p.m. +51 m.

h. m. SUN 1883 May 6 9 45 p.m. Visible, Philippines. (Mag. Total) 18 11 7 42 -- (Mag. Total) 1901 May 18 5 27 a.m. (civil time). -2 m.

h. m. MOON 1883 Oct. 16 6 54 a.m. Visible, California. (Mag. 0.28) 18 11 7 42 -- (Mag. 0.23) 1901 Oct. 27 2 36 p.m. -39 m.

h. m. SUN 1883 Oct. 30 11 37 p.m. Visible, N. Japan. (Mag. Annular) 18 11 7 42 -- (Mag. Annular) 1901 Nov. 11 7 19 a.m. (civil time) +1 m.

The foregoing does not by any means exhaust all that can be said respecting the Saros even on the popular side.

If the Saros comprised an exact number of days, each eclipse of a second Saros series would be visible in the same regions of the Earth as the corresponding eclipse in the previous series. But since there is a surplus fraction of nearly one-third of a day, each subsequent eclipse will be visible in another region of the Earth, which will be roughly a third of the Earth's circumference in longitude backwards (i.e. about 120?to the W.), because the Earth itself will be turned on its axis one-third forwards.

After what has been said as to the Saros and its use it might be supposed that a correct list of eclipses for 18.03 years would suffice for all ordinary purposes of eclipse prediction, and that the sequence of eclipses at any future time might be ascertained by adding to some one eclipse which had already happened so many Saros periods as might embrace the years future whose eclipses it was desired to study. This would be true in a sense, but would not be literally and effectively true, because corresponding eclipses do not recur exactly under the same conditions, for there are small residual discrepancies in the times and circumstances affecting the real movements of the Earth and Moon and the apparent movement of the Sun which, in the lapse of years and centuries, accumulate sufficiently to dislocate what otherwise would be exact coincidences. Thus an eclipse of the Moon which in A.D. 565 was of 6 digits[7] was in 583 of 7 digits, and in 601 nearly 8. In 908 the eclipse became total, and remained so for about twelve periods, or until 1088. This eclipse continued to diminish until the beginning of the 15th century, when it disappeared in 1413. Let us take now the life of an eclipse of the Sun. One appeared at the North Pole in June A.D. 1295, and showed itself more and more towards the S. at each subsequent period. On August 27, 1367, it made its first appearance in the North of Europe; in 1439 it was visible all over Europe; in 1601, being its 19th appearance, it was central and annular in England; on May 5, 1818, it was visible in London, and again on May 15, 1836. Its three next appearances were on May 26, 1854, June 6, 1872, and June 17, 1890. At its 39th appearance, on August 10, 1980, the Moon's shadow will have passed the equator, and as the eclipse will take place nearly at midnight (Greenwich M.T.), the phenomenon will be invisible in Europe, Africa, and Asia. At every succeeding period the central line of the eclipse will lie more and more to the S., until finally, on September 30, 2665, which will be its 78th appearance, it will vanish at the South Pole.[8]

The operation of the Saros effects which have been specified above, may be noticed in some of the groups of eclipses which have been much in evidence (as will appear from a subsequent chapter), during the second half of the 19th century. The following are two noteworthy Saros groups of Solar eclipses:--

1842 July 8. | 1850 Aug. 7. 1860 " 18. | 1868 " 17. 1878 " 29. | 1886 " 29. 1896 Aug. 9. | 1904 Sept. 9.

If the curious reader will trace, by means of the Nautical Almanac (or some other Almanac which deals with eclipses in adequate detail), the geographical distribution of the foregoing eclipses on the Earth's surface, he will see that they fulfil the statement made on p. 24 (ante), that a Saros eclipse when it reappears, does so in regions of the Earth averaging 120?of longitude to the W. of those in which it had, on the last preceding occasion, been seen; and also that it gradually works northwards or southwards.

But a given Saros eclipse in its successive reappearances undergoes other transformations besides that of Terrestrial longitude. These are well set forth by Professor Newcomb[9]:--

"Since every successive recurrence of such an eclipse throws the conjunction 28' further toward the W. of the node, the conjunction must, in process of time, take place so far back from the node that no eclipse will occur, and the series will end. For the same reason there must be a commencement to the series, the first eclipse being E. of the node. A new eclipse thus entering will at first be a very small one, but will be larger at every recurrence in each Saros. If it is an eclipse of the Moon, it will be total from its 13th until its 36th recurrence. There will be then about 13 partial eclipses, each of which will be smaller than the last, when they will fail entirely, the conjunction taking place so far from the node that the Moon does not touch the Earth's shadow. The whole interval of time over which a series of lunar eclipses thus extend will be about 48 periods, or 865 years. When a series of solar eclipses begins, the penumbra of the first will just graze the earth not far from one of the poles. There will then be, on the average, 11 or 12 partial eclipses of the Sun, each larger than the preceding one, occurring at regular intervals of one Saros. Then the central line, whether it be that of a total or annular eclipse, will begin to touch the Earth, and we shall have a series of 40 or 50 central eclipses. The central line will strike near one pole in the first part of the series; in the equatorial regions about the middle of the series, and will leave the Earth by the other pole at the end. Ten or twelve partial eclipses will follow, and this particular series will cease."

These facts deserve to be expanded a little.

We have seen that all eclipses may be grouped in a series, and that 18 years or thereabouts is the duration of each series, or Saros cycle. But these cycles are themselves subject to cycles, so that the Saros itself passes through a cycle of about 64 Saroses before the conditions under which any given start was made, come quite round again. Sixty-four times 18 make 1152, so that the duration of a Solar eclipse Great Cycle may be taken at about 1150 years. The progression of such a series across the face of the Earth is thus described by Mrs. Todd, who gives a very clear account of the matter:--

"The advent of a slight partial eclipse near either pole of the Earth will herald the beginning of the new series. At each succeeding return conformably to the Saros, the partial eclipse will move a little further towards the opposite pole, its magnitude gradually increasing for about 200 years, but during all this time only the lunar penumbra will impinge upon the Earth. But when the true shadow begins to touch, the obscuration will have become annular or total near the pole where it first appeared. The eclipse has now acquired a track, which will cross the Earth slightly farther from that pole every time it returns, for about 750 years. At the conclusion of this interval, the shadow path will have reached the opposite pole; the eclipse will then become partial again, and continue to grow smaller and smaller for about 200 years additional. The series then ceases to exist, its entire duration having been about 1150 years. The series of "great eclipses" of which two occurred in 1865 and 1883, while others will happen in 1901, 1919, 1937, 1955, and 1973, affords an excellent instance of the northward progression of eclipse tracks; and another series, with totality nearly as great (1850, 1868, 1886, 1904, and 1922), is progressing slowly southwards."

The word "Digit," formerly used in connection with eclipses, requires some explanation. The origin of the word is obvious enough, coming as it does from the Latin word Digitus, a finger. But as human beings have only eight fingers and two thumbs it is by no means clear how the word came to be used for twelfths of the disc of the Sun or Moon instead of tenths. However, such was the case; and when a 16th-century astronomer spoke of an eclipse of six digits, he meant that one-half of the luminary in question, be it Sun or Moon, was covered. The earliest use of the word "Digit" in this connection was to refer to the twelfth part of the visible surface of the Sun or Moon; but before the word went out of use, it came to be applied to twelfths of the visible diameter of the disc of the Sun or Moon, which was much more convenient. However, the word is now almost obsolete in both senses, and partial eclipses, alike of the Sun and of the Moon, are defined in decimal parts of the diameter of the luminary--tenths or hundredths according to the amount of precision which is aimed at. Where an eclipse of the Moon is described as being of more than 12 Digits or more than 1.0 (=? diameter) it is to be understood that the eclipse will be (or was) not only total, but that the Moon will be (or was) immersed in the Earth's shadow with a more or less considerable extent of shadow encompassing it.

There are some further matters which require to be mentioned connected with the periodicity of eclipses. To use a phrase which is often employed, there is such a thing as an "Eclipse Season," and what this is can only be adequately comprehended by looking through a catalogue of eclipses for a number of years arranged in a tabular form, and by collating the months or groups of months in which batches of eclipses occur. This is not an obvious matter to the casual purchaser of an almanac, who, feeling just a slight interest in the eclipses of a coming new year, dips into his new purchase to see what those eclipses will be. A haphazard glance at the almanacs of even two or three successive years will probably fail to bring home to him the idea that each year has its own eclipse season in which eclipses may occur, and that eclipses are not to be looked for save at two special epochs, which last about a month each, and which are separated from one another and from the eclipse seasons of the previous and of the following years by intervals of about six months, within a few days more or less. Such, however, is the case. A little thought will soon make it clear why such should be the case. We have already seen that the Moon's orbit, like that of every other planetary member of the Solar System, has two crossing places with respect to the ecliptic which are called "Nodes." We know also that the apparent motion of the Sun causes that body to traverse the whole of the ecliptic in the course of the year. The conjoint result of all this is that the Moon passes through a Node twice in every lunar month of 27 days, and the Sun passes (apparently) through a Node twice in every year. The first ultimate result of these facts is that eclipses can only take place at or near the nodal passages of the Moon and the Sun, and that as the Sun's nodal passages are separated by six months in every case the average interval between each set of eclipses, if there is more than one, must in all cases be six months, more or less by a few days, dependent upon the latitude and longitude of the Moon at or about the time of its Conjunction or Opposition under the circumstances already detailed. If the logic of this commends itself to the reader's mind, he will see at once why eclipses or groups of eclipses must be separated by intervals of about half an ordinary year. Hence it comes about that, taking one year with another, it may be said that we shall always have a couple of principal eclipses with an interval of half a year (say 183 days) between each; and that on either side of these dominant eclipses there will, or may be, a fortnight before or a fortnight after, two other pairs of eclipses with, in occasional years, one extra thrown in. It is in this way that we obtain what it has already been said dogmatically that we do obtain; namely, always in one year two eclipses, which must be both of the Sun, or any number of eclipses up to seven, which number will be unequally allotted to the Sun or to the Moon according to circumstances.

Though it is roughly correct to say that the two eclipse seasons of every year run to about a month each, in length, yet it may be desirable to be a little more precise, and to say that the limits of time for solar eclipses cover 36 days (namely 18 days before and 18 days after the Sun's nodal passages); whilst in the case of the Moon, the limits are the lesser interval of 23 days, being 11?on either side of the Moon's nodal passages.

We have already seen[10] that the Moon's nodes are perpetually undergoing a change of place. Were it not so, eclipses of the Sun and Moon would always happen year after year in the same pair of months for us on the Earth. But the operative effect of the shifting of the nodes is to displace backwards the eclipse seasons by about 20 days. For instance in 1899 the eclipse seasons fall in June and December. The middle of the eclipse seasons for the next succeeding 20 or 30 years will be found by taking the dates of June 8 and December 2, 1899, and working the months backwards by the amount of 19-2/3 days for each succeeding year. Thus the eclipse seasons in

1900 will fall in the months of May and November; accordingly amongst the eclipses of that year we shall find eclipses on May 28, June 13, and November 22.

Perhaps it would tend to the more complete elucidation of the facts stated in the last half dozen pages, if I were to set out in a tabular form all the eclipses of a succession, say of half a Saros or 9 years, and thus exhibit by an appeal to the eye directly the grouping of eclipse seasons the principles of which I have been endeavouring to define and explain in words.

Approximate Mid-interval.

1894. March 21. [Symbol: Moon] } March 29. * April 6. [Symbol: Sun] }

Sept. 15. [Symbol: Moon] } Sept. 22. ** Sept. 29. [Symbol: Sun] }

1895. March 11. [Symbol: Moon] } March 18. * March 26. [Symbol: Sun] }

Aug. 20. [Symbol: Sun] } Sept. 4. [Symbol: Moon] } Sept. 4. ** Sept. 18. [Symbol: Sun] }

1896. Feb. 13. [Symbol: Sun] } Feb. 20. * Feb. 28. [Symbol: Moon] }

Aug. 9. [Symbol: Sun] } Aug. 16. ** Aug. 23. [Symbol: Moon] }

1897. Feb. 1. [Symbol: Sun] Feb. 1. *

July 29. [Symbol: Sun] July 29. **

1898. Jan. 7. [Symbol: Moon] } Jan. 14. * Jan. 22. [Symbol: Sun] }

July 3. [Symbol: Moon] } July. 10. ** July 18. [Symbol: Sun] }

Dec. 13. [Symbol: Sun] } Dec. 27. [Symbol: Moon] } Dec. 27. * 1899. Jan. 11. [Symbol: Sun] }

June 8. [Symbol: Sun] } June 15. ** June 23. [Symbol: Moon] }

Dec. 2. [Symbol: Sun] } Dec. 9. * Dec. 16. [Symbol: Moon] }

1900. May 28. [Symbol: Sun] } June 5. ** June 13. [Symbol: Moon] }

Nov. 22. [Symbol: Sun] Nov. 22. *

1901. May 3. [Symbol: Moon] } May 10. ** May 18. [Symbol: Sun] }

Oct. 27. [Symbol: Moon] } Nov. 3. * Nov. 11. [Symbol: Sun] }

1902. April 8. [Symbol: Sun] } April 22. [Symbol: Moon] } April 22. ** May 7. [Symbol: Sun] }

Oct. 17. [Symbol: Moon] } Oct. 24. * Oct. 31. [Symbol: Sun] }

The Epochs in the last column which are marked with stars (*) or (**) as the case may be, represent corresponding nodes so that from any one single-star date to the next nearest single-star date means an interval of one year less (on an average) the 19-2/3 days spoken of on p. 32 (ante). A glance at each successive pair of dates will quickly disclose the periodical retrogradation of the eclipse epochs.

FOOTNOTES:

[Footnote 4: These limits are slightly different in the case of eclipses of the Moon. (See p. 190, post.)]

[Footnote 5: This assumes that 5 of these years are leap years.]

[Footnote 6: If there are 5 leap years in the 18, the odd days will be 10; if 4 they will be 11; if only 3 leap years (as from 1797 to 1815 and 1897 to 1915), the odd days to be added will be 12.]

[Footnote 7: See p. 28 (post) for an explanation of this word.]

[Footnote 8: In Mrs. D. Todd's interesting little book, Total Eclipses of the Sun (Boston, U.S., 1894), which will be several times referred to in this work, two maps will be found, which will help to illustrate the successive northerly or southerly progress of a series of Solar eclipses, during centuries.]

[Footnote 9: In his and Professor Holden's Astronomy for Schools and Colleges, p. 184.]

[Footnote 10: See p. 19 (ante).]

CHAPTER IV.

MISCELLANEOUS THEORETICAL MATTERS CONNECTED WITH ECLIPSES OF THE SUN (CHIEFLY).

One or two miscellaneous matters respecting eclipses of the Sun (chiefly) will be dealt with in this chapter. It is not easy to explain or define in words the circumstances which control the duration of a Solar eclipse, whereas in the case of a lunar eclipse the obscuration is the same in degree at all parts of the Earth where the Moon is visible. In the case of a Solar eclipse it may be total, perhaps, in Africa, may be of six digits only in Spain, and of two only in England. Under the most favourable circumstances the breadth of the track of totality across the Earth cannot be more than 170 miles, and it may be anything less than that down to zero where the eclipse will cease to be total at all, and will become annular. The question whether a given eclipse shall exhibit itself on its central line as a total or an annular one depends, as has been already explained, on the varying distances of the Earth and the Moon from the Sun in different parts of their respective orbits. Hence it follows that not only may an eclipse show itself for several Saros appearances as total and afterwards become annular, and vice versa but on rare occasions one and the same eclipse may be annular in one part of its track across the Earth and total in another part, a short time earlier or later. This last-named condition might arise because the Moon's distance from the Earth or the Sun had varied sufficiently during the progress of the eclipse to bring about such a result; or because the shadow just reaching the Earth and no more the eclipse would be total only for the moment when a view perpendicular upwards could be had of it, and would be annular for the minutes preceding and the minutes following the perpendicular glimpse obtained by observers actually on the central line. The eclipse of December 12, 1890, was an instance of this.

If the paths of several central eclipses of the Sun are compared by placing side by side a series of charts, such as those given in the Nautical Almanac or in Oppolzer's Canon, it will be noticed that the direction of the central line varies with the season of the year. In the month of March the line runs from S.W. to N.E., and in September from N.W. to S.E. In June the line is a curve, going first to the N.E. and then to the S.E. In December the state of things is reversed, the curve going first to the S.E. and then to the N.E. At all places within about 2000 miles of the central line the eclipse will be visible, and the nearer a place is to the central line, so much the larger will be the portion of the Sun's disc concealed from observers there by the Moon. If the central line runs but a little to the N. of the Equator in Winter or of 25?of N. latitude in Summer, the eclipse will be visible all over the Northern Hemisphere, and the converse will apply to the Southern Hemisphere. It is something like a general rule in the case of total and annular eclipses, though subject to many modifications, that places within 200-250 miles of the central line will have partial eclipse of 11 digits; from thence to 500 miles of 10 digits, and so on, diminishing something like 1 digit for every 250 miles, so that at 2000 miles, or rather more, the Sun will be only to a very slight extent eclipsed, or will escape eclipse altogether.

The diameter of the Sun being 866,000 miles and the Moon being only 2160 miles or 1/400th how comes it to be possible that such a tiny object should be capable of concealing a globe 400 times bigger than itself? The answer is-- Distance. The increased distance does it. The Moon at its normal distance from the Earth of 237,000 miles could only conceal by eclipse a body of its own size or smaller, but the Sun being 93,000,000 miles away, or 392 times the distance of the Moon, the fraction 1/392 representing the main distance of the Moon, more than wipes out the fraction 1/400 which represents our satellite's smaller size.

During a total eclipse of the Sun, the Moon's shadow travels across the Earth at a prodigious pace--1830 miles an hour; 30?miles a minute; or rather more than a ?mile a second. This great velocity is at once a clue to the fact that the total phase during an eclipse of the Sun lasts for so brief a time as a few minutes; and also to the fact that the shadow comes and goes almost without being seen unless a very sharp watch is kept for it. Indeed, it is only observers posted on high ground with some miles of open low ground spread out under their eyes who have much chance of detecting the shadow come up, go over them, and pass forwards.

Places at or near the Earth's equator enjoy the best opportunities for seeing total eclipses of the Sun, because whilst the Moon's shadow travels eastwards along the Earth's surface at something like 2000 miles an hour, an observer at the equator is carried in the same direction by virtue of the Earth's axial rotation at the rate of 1040 miles an hour. But the speed imparted to an observer as the result of the Earth's axial rotation diminishes from the equator towards the poles where it is nil, so that the nearer he is to a pole the slower he goes, and therefore the sooner will the Moon's shadow overtake and pass him, and the less the time at his disposal for seeing the Sun hidden by the Moon.

It was calculated by Du Sour that the greatest possible duration of the total phase of a Solar eclipse at the equator would be 7^m 58^s, and for the latitude of Paris 6^m 10^s. In the case of an annular eclipse the figures would be 12^m 24^s for the equator, and 9^m 56^s for the latitude of Paris. These figures contemplate a combination of all the most favourable circumstances possible; as a matter of fact, I believe that the greatest length of total phase which has been actually known was 6m and that was in the case of the eclipse of August 29, 1886. It was in the open Atlantic that this duration occurred, but it was not observed. The maximum observed obscuration during this eclipse was no more than 4^m.

Though total eclipses of the Sun happen with tolerable frequency so far as regards the Earth as a whole, yet they are exceedingly rare at any given place. Take London, for instance. From the calculations of Hind, confirmed by Maguire,[11] it may be considered as an established fact that there was no total eclipse visible at London between A.D. 878 and 1715, an interval of 837 years. The next one visible at London, though uncertain, is also a very long way off. There will be a total eclipse on August 11, 1999, which will come as near to London as the Isle of Wight, but Hind, writing in 1871, said that he doubted whether there would be any other total eclipse "visible in England for 250 years[12] from the present time." Maguire states that the Sun has been eclipsed, besides twice at London, also twice at Dublin, and no fewer than five times at Edinburgh during the 846 years examined by him. In fact that every part of the British Isles has seen a total eclipse at some time or other between A.D. 878 and 1724 except a small tract of country at Dingle, on the West coast of Ireland. The longest totality was on June 15, 885, namely,

4^m 55^s, and the shortest in July 16, 1330, namely, 0^m 56^s.

Contrast with this the obscure island of Blanquilla, off the northern coast of Venezuela. The inhabitants of that island not long ago had the choice of two total eclipses within three and a half years, namely, August 29, 1886, and December 22, 1889; whilst Yellowstone, U.S., had two in twelve years (July 29, 1878, and January 1, 1889).

Counting from first to last, Du Sour found that at the equator an eclipse of the Sun might last 4^h 29^m, and at the latitude of Paris 3^h 26^m. These intervals, of course, cover all the subordinate phases. The total phase which alone (with perhaps a couple of minutes added) is productive of spectacular effects, and interesting scientific results is a mere matter of minutes which may be as few as one (or less), or only as many as 6 or 8.

As a rule, a summer eclipse will last longer than a winter one, because in summer the Earth (and the Moon with it), being at its maximum distance from the Sun, the Sun will be at its minimum apparent size, and therefore the Moon will be able to conceal it the longer.

FOOTNOTES:

[Footnote 11: Month. Not., R.A.S., vol. xlv., p. 400. June 1885.]

[Footnote 12: Johnson makes the eclipse of June 14, 2151, to be "nearly, if not quite, total at London." Possibly it was this eclipse which Hind had in his thoughts when he wrote in the Times (July 28, 1871) the passage quoted above.]

CHAPTER V.

WHAT IS OBSERVED DURING THE EARLIER STAGES OF AN ECLIPSE OF THE SUN.

The information to be given in this and the next following chapters will almost exclusively concern total and annular eclipses of the Sun, for, in real truth, there is practically only one thing to think about during a partial eclipse of the Sun. This is, to watch when the Moon's black body comes on to the Sun and goes off again, for there are no subsidiary phenomena, either interesting or uninteresting, unless, indeed, the eclipse should be nearly total. The progress of astronomical science in regard to eclipses has been so extensive and remarkable of late years that, unless the various points for consideration are kept together under well-defined heads, it will be almost impossible either for a writer or a reader to do full justice to the subject. Having regard to the fact that the original conception of this volume was that it should serve as a forerunner to the total solar eclipse of May 28, 1900 (and through that to other total eclipses), from a popular rather than from a technical standpoint, I think it will be best to mention one by one the principal features which spectators should look out for, and to do so as nearly as may be in the order which Nature itself will observe when the time comes.

Of course the commencement of an eclipse, which is virtually the moment when the encroachment on the circular outline of the Sun by the Moon begins, or can be seen, though interesting as a proof that the astronomer's prophecy is about to be fulfilled, is not a matter of any special importance, even in a popular sense, much less in a scientific sense. As a rule, the total phase does not become imminent, so to speak, until a whole hour and more has elapsed since the first contact; and that hour will be employed by the scientific observer, less in looking at the Sun than in looking at his instruments and apparatus. He will do this for the purpose of making quite sure that everything will be ready for the full utilisation to the utmost extent of the precious seconds of time into which all his delicate observations have to be squeezed during the total phase.

With these preliminary observations I shall proceed now to break up the remainder of what I have to say respecting total eclipses into what suggest themselves as convenient sectional heads.

THE MOON'S SHADOW AND THE DARKNESS IT CAUSES.

In awaiting the darkness which is expected to manifest itself an unthinking and inexperienced observer is apt to look out for the coming obscurity, as he looks out for night-fall half an hour or more after sunset and during the evening twilight. The darkness of an eclipse is all this and something more. It is something more in two senses; for the interval of time between the commencement of an eclipse and totality is different in duration and different in quality, so to speak, from the diminution of daylight on the Earth which ensues as the twilight of evening runs its course. Speaking roughly, sunset may be described as an almost instantaneous loss of full sunlight; and the gradual loss of daylight is noticeable even at such short intervals as from one five minutes to another. This is by no means the case previous to a total eclipse of the Sun. When that is about to occur, the reduction of the effective sunlight is far more gradual. For instance, half an hour after an eclipse has commenced more than half the Sun's disc will still be imparting light to the Earth: but half an hour after sunset the deficiency of daylight will be very much more marked and, if no artificial light is at hand, very much more inconvenient.

If there should be within easy reach of the observer's post a bushy tree, such for instance as an elm, 30 ft. or 40 ft. high, and spreading out sufficiently for him to place himself under it in a straight line with the Sun, and with a nice smooth surface of ground for the sun's rays to fall on, he will see a multitude of images of the Sun thrown upon the ground.

Until the eclipse has commenced these images will be tiny circles overlapping one another, and of course each of these circles means so many images of the Sun. These images indeed can be seen on any fine day, and the circles increase in size in proportion to the height of the foliage above the ground, being something like 1 inch for every 10 feet. It may be remarked, by the way, that the images are circles, because the Sun is a source of light having a circular outline, and is not a point of light like a star. If it were, the outline of the foliage would be reproduced on the ground leaf for leaf. It follows naturally from all this that when in consequence of there being an eclipse in progress the shape of the Sun's contour gradually changes, so will the shape of the Solar images on the ground change, becoming eventually so many crescents. Moreover, the horns of the crescent-shaped images will be in the reverse direction to the horns of the actual crescent of the Sun at the moment, the rays of the Sun crossing as they pass through the foliage, just as if each interstice were a lens.

Supposing there are some spots on the Sun at a time when an eclipse is in progress the Moon's passage over these spots may as well be noticed. In bygone years some amount of attention was devoted to this matter with the view of ascertaining whether any alteration took place in the appearance of the spots; distortion, for instance, such as might be produced by the intervention of a lunar atmosphere. No such distortion was ever noticed, and observations with this idea in view may be said to possess now only an academic interest, for it may be regarded as a well-established fact that the Moon has no atmosphere.

During the passage of the Moon over Sun-spots an opportunity is afforded of comparing the blackness, or perhaps we should rather say, the intensity of the shade of a Sun-spot with the blackness of the Moon's disc. Testimony herein is unanimous that the blackness of the Moon during the stages of partial eclipse is intense compared with the darkest parts of a Sun-spot; and this, be it remembered, in spite of the fact that during the partial phase the atmosphere between the observer and the Sun is brilliantly illuminated, whilst the Moon itself, being exposed to Earth-shine, is by no means absolutely devoid of all illumination.

When the Moon is passing across the Sun there have often been noticed along the limb of the Moon fringes of colour, and dark and bright bands. This might not necessarily be a real appearance for it is conceivable that such traces of colour might be due to the telescopes employed not having been truly achromatic, that is, not sufficiently corrected for colour; but making every allowance for this possible source of mistake there yet remains proof that the colour which has often been seen has been real.

As to whether the Moon's limb can be seen during a partial eclipse, or during the partial phase of what is to be a total eclipse, the evidence is somewhat conflicting. There is no doubt that when the totality is close at hand the Moon's limb can be seen projected on the Corona (presently to be described); but the question is, whether the far-off limb of the Moon can be detected in the open sky whilst something like full daylight still prevails on the Earth. Undoubtedly the preponderance of evidence is against the visibility of the Moon as a whole, under such circumstances; but there is nevertheless some testimony to the contrary. A French observer, E. Liais, said that three photographic plates of the eclipse of 1858 seen in S. America all showed the outer limb of the Moon with more or less distinctness. This testimony, be it noted, is photographic and not visual; and on the whole it seems safest to say that there is very small probability of the Moon as a whole ever being seen under the circumstances in question.

What has just been said concerns the visibility of the Moon during quite the early, or on the other hand during quite the late, stages of a total eclipse. Immediately before or after totality the visibility of the whole contour of the Moon is a certain fact; and the only point upon which there is a difference of opinion is as to what are the time-limits beyond which the Moon must not be expected to be seen. The various records are exceedingly contradictory: perhaps the utmost that can be said is that the whole Moon must not be expected to be visible till about 20 minutes before totality, or for more than 5 minutes after totality--but it must be admitted that these figures are very uncertain in regard to any particular eclipse.

It has been sometimes noticed when the crescent of the Sun had become comparatively small, say that the Sun was about 7/8ths covered, that the uncovered portion exhibited evident colour which has been variously described as "violet," "brick-red," "reddish," "pink," "orange," "yellowish." The observations on this point are not very numerous and, as will appear from the statement just made, are not very consistent; still it seems safe to assume that a hue, more or less reddish, does often pervade the uncovered portion of a partially-eclipsed Sun.

The remark just made as regards the Sun seems to have some application to the Moon. There are a certain number of instances on record that what is commonly spoken of as the black body of the Moon does, under certain circumstances, display traces of red which has been variously spoken of as "crimson," "dull coppery," "reddish-brownish" and "dull glowing coal."

SHADOW BANDS.

Let us suppose that we have a chance of observing a total eclipse of the Sun; have completed all our preliminary preparations; have taken note of everything which needs to be noted or suggests itself for that purpose up till nearly the grand climax; and that the clock tells us that we are within, say, five minutes of totality. Somewhere about this time perhaps we shall be able to detect, dancing across the landscape, singular wavy lines of light and shade. These are the "Shadow Bands," as they are called. The phrase is curiously inexplicit, but seemingly cannot be improved upon at present because the philosophy of these appearances--their origin and the laws which regulate their visibility--are unknown, perhaps because amid the multitude of other things to think about sufficient attention has hitherto not been paid to the study of them. These shadow bands are most striking if a high plastered wall, such as the front of a stone or stuccoed house, is in their track as a screen to receive them. The shadow bands seem to vary both in breadth and distance apart at different eclipses, and also in the speed with which they pass along. Though, as already stated, little is known of their origin yet they may be conceived to be due to irregularities in the atmospheric refraction of the slender beam of light coming from the waning or the waxing crescent of the Sun, for be it understood they may be visible after totality as well as before it. It is to be remarked that they have never been photographed.

In addition to the shadow bands there are instances on record of the limbs of the Sun's crescent appearing to undulate violently on the approach of totality. These undulations were noticed by Airy in 1842 about 6 minutes before totality. Blake, in America in 1869, observed the same phenomenon 8 minutes before totality. In other cases the interval would seem to have been very much shorter--a mere matter of seconds. A very singular observation was made in 1858 by Mr. J.爨 Smith at Laycock Abbey, Wiltshire, on the occasion of the annular eclipse of that year. He says[13]:--"Both my brother and myself were distinctly impressed with the conviction that the withdrawal of light was not continuous, but by pulsations, or, as it were, waves of obscuration, the darkness increasing by strokes which sensibly smote the eye, and were repeated distinctly some five or seven times after we had remarked the phenomenon and before the time of greatest obscuration. This did not occur on the return of light, which came back continuously and without shock or break." Roker mentions that though this phenomenon was very apparent to the naked eye it was not visible in the telescope.

Faint rays or brushes of light sometimes seem to spring from the diminishing crescent of the Sun. These rays generally are very transient and not very conspicuous, and perhaps must be distinguished as regards both their appearance and their origin from the more striking rays which are usually seen a few minutes before or after totality, and which are generally associated with, or even deemed to belong to, the Corona. Fig. 7 represents these rays as seen in Spain on July 18, 1860, some minutes after totality. They are described as having been well defined, but at some moments more marked than at others, and though well-defined yet constantly varying.

Radiations of light more or less of the character just described may probably be regarded as a standing feature of every total eclipse.

THE APPROACH OF TOTALITY.

The next thing to think about and to look out for is the approach of the Moon's shadow. I have mentioned this already,[14] and also the appalling velocity with which it seems to approach. By this time the coming darkness, which characterises every total phase, will have reached an advanced stage of development. The darkness begins to be felt. The events which manifest themselves at this juncture on the Earth (rather than in the sky around the Sun) are so graphically described by the American writer whom I have already quoted, and who writes, moreover, from personal experience, that I cannot do better than transfer her striking account to my pages.[15] "Then, with frightful velocity, the actual shadow of the Moon is often seen approaching, a tangible darkness advancing almost like a wall, swift as imagination, silent as doom. The immensity of nature never comes quite so near as then, and strong must be the nerves not to quiver as this blue-black shadow rushes upon the spectator with incredible speed. A vast, palpable presence seems overwhelming the world. The blue sky changes to gray or dull purple, speedily becoming more dusky, and a death-like trance seizes upon everything earthly. Birds, with terrified cries, fly bewildered for a moment, and then silently seek their night-quarters. Bats emerge stealthily. Sensitive flowers, the scarlet pimpernel, the African mimosa, close their delicate petals, and a sense of hushed expectancy deepens with the darkness. An assembled crowd is awed into absolute silence almost invariably. Trivial chatter and senseless joking cease. Sometimes the shadow engulfs the observer smoothly, sometimes apparently with jerks; but all the world might well be dead and cold and turned to ashes. Often the very air seems to hold its breath for sympathy; at other times a lull suddenly awakens into a strange wind, blowing with unnatural effect. Then out upon the darkness, gruesome but sublime, flashes the glory of the incomparable corona, a silvery, soft, unearthly light, with radiant streamers, stretching at times millions of uncomprehended miles into space, while the rosy, flaming protuberances skirt the black rim of the Moon in ethereal splendour. It becomes curiously cold, dew frequently forms, and the chill is perhaps mental as well as physical. Suddenly, instantaneous as a lightning flash, an arrow of actual sunlight strikes the landscape, and Earth comes to life again, while corona and protuberances melt into the returning brilliance, and occasionally the receding lunar shadow is glimpsed as it flies away with the tremendous speed of its approach."

In connection with the approach of the Moon's shadow, it is to be noted that at totality the heavens appear in a certain sense to descend upon the Earth. If an observer is looking upwards towards the zenith over his head, he will see the clouds appear to drop towards the Earth, and the surrounding gloom seems also to have the effect of vitiating one's estimate of distances. To an observer upon a high hill, a plain below him appears to become more distant. Although what has been called the descent of the clouds (that is to say their appearance of growing proximity) is most manifest immediately before the totality, yet a sense of growing nearness may sometimes be noticed a very considerable time before the total phase is reached.

Whilst on the subject of clouds, it may be mentioned that although there is in the vault of heaven generally during the total phase an appreciable sensation of black darkness, more or less absolute, that is to say, either blackish or greyish, yet in certain regions of the sky, (generally in the direction of the horizon) the clouds, when there are any, often exhibit colours in strata, orange hue below and red above, with indigo or grey or black higher up still, right away to the Sun's place. The cause of these differences is to be found in the fact that the lower part of the atmosphere within the area of the Moon's shadow is, under the circumstances in question, illuminated by light which having passed through many miles of atmosphere near to the Earth's surface, has lost much from the violet end of its spectrum, leaving an undue proportion of the red end.

On certain occasions iridescent or rainbow-tinted clouds may be seen in the vicinity of the Sun, either before, or during, or after totality, depending on circumstances unknown. Such clouds have been observed at all these three stages of a total eclipse. The effects of course are atmospheric, and have no physical connection with either Sun or Moon.

THE DARKNESS OF TOTALITY.

With respect to the general darkness which prevails during totality, great discrepancies appear in the accounts, not only as between different eclipses, but in respect of the same eclipse observed by different people at different places. Perhaps the commonest test applied by most observers is that of the facility or difficulty of reading the faces of chronometers or watches. Sometimes this is done readily, at other times with difficulty. In India in 1868, one observer stated that it was impossible to recognise a person's face three yards off, and lamplight was needed for reading his chronometer. On the other hand in Spain in 1860, it was noted that a thermometer, as well as the finest hand-writing, could be read easily. The foregoing remarks apply to the state of things in the open air. In 1860, it was stated that inside a house in Spain the darkness was so great that people moving about had to take great care lest they should run violently against the household furniture.

Perhaps on the whole it may be said that the darkness of an ordinary totality is decidedly greater than that of a full Moon night.

Many observers have noted during totality that even when there has not been any very extreme amount of absolute darkness, yet the ruddy light already mentioned as prevailing towards the horizon often gives rise to weird unearthly effects, so that the faces of bystanders assume a sickly livid hue not unlike that which results from the light of burning salt.

METEOROLOGICAL AND OTHER EFFECTS.

It is very generally noticed that great changes take place in the meteorological conditions of the atmosphere as an eclipse of the Sun runs its course from partial phase to totality, and back again to partial phase. It goes without saying that the obstruction of the solar rays by the oncoming Moon would necessarily lead to a steady and considerable diminution in the general temperature of the air. This has often been made the matter of exact thermometric record, but it is not equally obvious why marked changes in the wind should take place. As the partial phase proceeds it is very usual for the wind to rise or blow in gusts and to die away during totality, though there are many exceptions to this, and it can hardly be called a rule.

The depression of temperature varies very much indeed according to the locality where the eclipse is being observed and the local thermometric conditions which usually prevail. The actual depression will often amount to 10?or 20?and the deposit of dew is occasionally noticed.

In addition to the general effects of a total solar eclipse on men, animals, and plants as summarised in the extract already made from Mrs. Todd's book a few additional particulars may be given culled from many recorded observations. Flowers and leaves which ordinarily close at night begin long before totality to show signs of closing up. Thus we are told that in 1836 "the crocus, gentian and anemone partially closed their flowers and reopened them as the phenomenon passed off: and a delicate South African mimosa which we had reared from a seed entirely folded its pinnate leaves until the Sun was uncovered." In 1851 "the night violet, which shortly before the beginning of the eclipse had little of its agreeable scent about it, smelt strongly during the totality."

In the insect world ants have been noticed to go on working during totality, whilst grasshoppers are stilled by the darkness, and earth-worms come to the surface. Birds of all kinds seem always upset in their habits, almost invariably going to roost as the darkness becomes intensified before totality. In 1868 "a small cock which had beforehand been actively employed in grubbing about in the sand went to sleep with his head under his wing and slept for about 10 minutes, and on waking uttered an expression of surprise, but did not crow." In 1869 mention is made of an unruly cow "accustomed to jump into a corn-field at night" being found to have trespassed into the said corn-field during the total phase.

The thrilling descriptions of the effects of the oncoming darkness of totality, derived from the records of past total eclipses, are not likely to be improved upon in the future, for we shall receive them more and more from amateurs and less and less from astronomical experts. Every additional total eclipse which happens testifies to the fact that the time and thoughts of these latter classes of people will be to an increasing degree dedicated to instrumental work rather than to simple naked eye or even telescopic observation. The spectroscope and the camera are steadily ousting the simple telescope of every sort and unassisted eye observations from solar eclipse work.

Mrs. Todd has the following apt remarks by way of summary of the results to an individual of observing a total eclipse of the Sun:--"I doubt if the effect of witnessing a total eclipse ever quite passes away. The impression is singularly vivid and quieting for days, and can never be wholly lost. A startling nearness to the gigantic forces of Nature and their inconceivable operation seems to have been established. Personalities and towns and cities, and hates and jealousies, and even mundane hopes, grow very small and very far away."

FOOTNOTES:

[Footnote 13: Month. Not., R.A.S., vol. xviii. p. 251.]

[Footnote 14: See p. 36 (ante).]

[Footnote 15: Mrs. D. Todd, Total Eclipses of the Sun, p. 21.]

CHAPTER VI.

WHAT IS OBSERVED DURING THE TOTAL PHASE OF AN ECLIPSE OF THE SUN.

The central feature of every total eclipse of the Sun is undoubtedly the Corona[16] and the phenomena connected with it; but immediately before the extinction of the Sun's light and incidental thereto there are some minor features which must be briefly noticed.

[Illustration: FIG. 8.--BRUSHES OF LIGHT.]

The Corona first makes its appearance on the side of the dark Moon opposite to the disappearing crescent, but brushes of light are sometimes observed on the same side, along the convex limb of the disappearing crescent. The appearance of the brushes will be sufficiently realised by an inspection of the annexed engraving without the necessity of any further verbal description. These brushes are little, if at all, coloured, and must not be confused with the "Red Flames" or "Prominences" hereafter to be described.

BAILY'S BEADS.

When the disc of the Moon has advanced so much over that of the Sun as to have reduced the Sun almost to the narrowest possible crescent of light, it is generally noticed that at a certain stage the crescent suddenly breaks up into a succession of spots of light. These spots are sometimes spoken of as "rounded" spots, but it is very doubtful whether (certainly in view of their supposed cause) they could possibly be deemed ever to possess an outline, which by any stretch, could be called "rounded." Collating the recorded descriptions, some such phrase as "shapeless beads" of light would seem to be the most suitable designation. These are observed to form before the total phase, and often also after the total phase has passed. Under the latter circumstances, the beads of light eventually run one into another, like so many small drops of water merging into one big one. The commonly received explanation of "Baily's Beads" is that they are no more than portions of the Sun's disc, seen through valleys between mountains of the Moon, the said mountains being the cause why the bright patches are discontinuous. It is exceedingly doubtful whether this is the true explanation. The whole question is involved in great uncertainty, and well deserves careful study during future eclipses; but this it is not likely to get, in view of the current fashion of every sufficiently skilled observer concentrating his attention on matters connected with the solar Corona (observed spectroscopically or otherwise), to the exclusion of what may be called older subjects of study. I will dismiss Baily's Beads from our consideration with the remark that the first photograph of them was obtained at Ottumwa, Illinois, U.S., during the eclipse of 1869.

"Baily's Beads" received their name from Mr. Francis Baily, who, in 1836, for the first time exhaustively described them; but they were probably seen and even mentioned long before his time. At the total eclipse of the Sun, seen at Penobscot in North America, on October 27, 1780, they would seem to have been noticed, and perhaps even earlier than that date.

Almost coincident with the appearance of Baily's Beads, that is, either just before or just after, and also just before or just after the absolute totality (there seems no certain rule of time) jets of red flame are seen to dart out from behind the disc of the Moon. It is now quite recognised as a certain fact that these "Red Flames" belong to the Sun and are outbursts of hydrogen gas. Moreover, they are now commonly called "Prominences," and with the improved methods of modern science may be seen almost at any time when the Sun is suitably approached; and they are not restricted in their appearance to the time when the Sun is totally eclipsed as was long supposed.

I may have more to say about these Red Flames later on; but am at present dealing only with the outward appearances of things. Carrington's description has been considered very apt. One which he saw in 1851 he likened to "a mighty flame bursting through the roof of a house and blown by a strong wind."

Certain ambiguous phrases made use of in connection with eclipses of ancient date may perhaps in reality have been allusions to the Red Flames; otherwise the first account of them given with anything like scientific precision seems to be due to a Captain Stannyan, who observed them at Berne during the eclipse of 1706. His words are that the Sun at "his getting out of his eclipse was preceded by a blood-red streak from its left limb which continued not longer than six or seven seconds of time; then part of the Sun's disc appeared all of a sudden."

Some subsequent observers spoke of the Red Flames as isolated jets of red light appearing here and there; whilst others seem to have thought they had seen an almost or quite continuous ring of red light around the Sun. The last-named idea is now recognised as the more accurate representation of the actual facts, the Red Flames being emanations proceeding from a sort of shell enveloping the Sun, to which shell the name of "Chromosphere" has now come to be applied.

As regards the Moon itself during the continuance of the total phase, all that need be said is that our satellite usually exhibits a disc which is simply black; but on occasions observers have called it purple or purplish. Although during totality the Moon is illuminated by a full allowance of Earth-shine (light reflected by the Earth into space), yet from all accounts this is always insufficient to reveal any traces of the irregularities of mountains and valleys, etc., which exist on the Moon.

When during totality any of the brighter planets, such as Mercury, Venus, Mars, Jupiter, or Saturn, happen to be in the vicinity of the Sun they are generally recognised; but the stars seen are usually very few, and they are only very bright ones of the 1st or 2nd magnitudes. Perhaps an explanation of the paucity of stars noticed is to be found in the fact that the minds of observers are usually too much concentrated on the Sun and Moon for any thought to be given to other things or other parts of the sky.

Perhaps this is a convenient place in which to recall the fact that there has been much controversy in the astronomical world during the last 50 years as to whether there exist any undiscovered planets revolving round the Sun within the orbit of Mercury. Whilst there is some evidence, though slight, that one or more such planets have been seen, opponents of the idea base their scepticism on the fact that with so many total eclipses as there have been since 1859 (when Lescarbault claimed to have found a planet which has been called "Vulcan"), no certain proof has been obtained of the existence of such a planet; and what better occasion for finding one (if one exists of any size) than the darkness of a total solar eclipse? At present it must be confessed that the sceptics have the best of it.

THE CORONA.

We have now to consider what I have already called the central feature of every total eclipse. It was long ago compared to the nimbus often placed by painters around the heads of the Virgin Mary and other saints of old; and as conveying a rough general idea the comparison may still stand. It has been suggested that not a bad idea of it may be obtained by looking at a Full Moon through a wire-gauze window-screen. The Corona comes into view a short time (usually to be measured by seconds) before the total extinction of the Sun's rays, lasts during totality and endures for a brief interval of seconds (or it might be a minute) after the Sun has reappeared. It was long a matter of discussion whether the Corona belonged to the Sun or the Moon. In the early days of telescopic astronomy there was something to be said perhaps on both sides, but it is now a matter of absolute certainty that it belongs to the Sun, and that the Moon contributes nothing to the spectacle of a total eclipse of the Sun, except its own solid body, which blocks out the Sun's light, and its shadow, which passes across the Earth.

Of the general appearance of the Corona some idea may be obtained from Fig. 1 (see Frontispiece) which so far as it goes needs little or no verbal description. Stress must however be laid on the word "general" because every Corona may be said to differ from its immediate predecessor and successor, although, as we shall see presently, there is strong reason to believe that there is a periodicity in connection with Coronas as with so many other things in the world of Astronomy. A curious point may here be mentioned as apparently well established, namely, that when long rays are noticed in the Corona they do not seem to radiate from the Sun's centre as the short rays more or less seem to do. Though the aggregate brilliancy of the Corona varies somewhat yet it may be taken to be much about equal on the whole to the Moon at its full. The Corona is quite unlike the Moon as regards heat for its radiant heat has been found to be very well marked.

There is another thing connected with the Sun's Corona which needs to be mentioned at the outset and which also furnishes a reason for treating it in a somewhat special manner. The usual practice in writing about science is to deal with it in the first instance descriptively, and then if any historical information is to be given to exhibit that separately and subsequently. But our knowledge of the Sun's Corona has developed so entirely by steps from a small beginning that it is neither easy nor advantageous to keep the history separate or in the background and I shall therefore not attempt to do so.

Astronomers are not agreed as to what is the first record of the Corona. It is commonly associated with a total eclipse which occurred in the 1st century A.D. and possibly in the year 96 A.D. Some details of the discussion will be found in a later chapter,[17] and I will make no further allusion to the matter here. Passing over the eclipses of 968 A.D. and 1030 A.D. the records of both of which possibly imply that the Corona was noticed, we may find ourselves on thoroughly firm ground in considering the eclipse of April 9, 1567. Clavius, a well-known writer on chronology, undoubtedly saw then the Corona in the modern acceptation of the word but thought it merely the uncovered rim of the Sun. In reply to this Kepler showed by some computations of his own, based on the relative apparent sizes of the Sun and Moon, that Clavius's theory was untenable. Kepler, however, put forth a theory of his own which was no better, namely, that the Corona was due to the existence of an atmosphere round the Moon and proved its existence. From this time forwards we have statements, by various observers, applying to various eclipses, of the Corona seeming to be endued with a rotatory motion. The Spanish observer, Don A. Ulloa, in 1778, wrote thus respecting the Corona seen in that year:--"After the immersion we began to observe round the Moon a very brilliant circle of light which seemed to have a rapid circular motion something similar to that of a rocket turning about its centre." Modern observations furnish no counterpart of these ideas of motion in the Corona. Passing over many intervening eclipses we must note that of 1836 (which gave us "Baily's Beads") as the first which set men thinking that total eclipses of the Sun exhibited subsidiary phenomena deserving of careful and patient attention. Such attention was given on the occasion of the eclipses of 1842 and 1851, still however without the Corona attracting that interest which it has gained for itself more recently. It was noticed indeed that the Corona always first showed itself on the side of the Moon farthest from the vanishing crescent but the full significance of this fact was not at first realised. Mrs. Todd well remarks:--"In the early observations of the Corona it was regarded as a halo merely and so drawn. Its real structure was neither known, depicted, nor investigated. The earliest pictures all show this. Preconceived ideas prejudiced the observers, and their sketches were mostly structureless.... It should not be forgotten that the Coronal rays project outward into space from a spherical Sun and do not lie in a plane as they appear to the eye in photographs and drawings." After remarking on the value of photographs of the Corona up to a certain point because of their automatic accuracy Mrs. Todd very sensibly says, "but pencil drawings, while ordinarily less trustworthy because involving the uncertain element of personal equation are more valuable in delineating the finest and faintest detail of which the sensitive plate rarely takes note; the vast array of both, however, shows marked differences in the structure and form of the Corona from one eclipse to another though it has not yet revealed rapid changes during any one observation. This last interesting feature can be studied only by comparison of photographs near the beginning of an eclipse track and its end, two or three hours of absolute time apart." Concerted efforts to accomplish this were made in 1871, 1887, and 1889, but they broke down because the weather failed at one or other end of the chain of observing stations and a succession of photographs not simultaneous but separated by sufficient intervals of time could not be had. The eclipse of 1893, however, yielded successful though negative results. Photographs in South America compared with photographs in Africa two hours later in time disclosed no appreciable difference in the structure of the Corona and its streamers. The eclipse of May 28, 1900, will furnish the next favourable opportunity for a repetition of this experiment by reason of the fact that the line of totality begins in North America, crosses Portugal and Spain and ceases in Africa. In other words, traverses countries eminently calculated to facilitate the establishment of photographic observing stations where observations can be made not simultaneously but at successive intervals spread over several hours.

Although of course the Corona had been observed long before the year 1851, as indeed we have already seen, yet the eclipse of 1851 is the farthest back which we can safely take as a starting-point for gathering up thoroughly precise details, because it was the first at which photography was brought into use. Starting, therefore, with that eclipse I want to lay before the reader some of the very interesting and remarkable generalisations which (thanks especially to Mr. W.歟. Wesley's skilful review of many of the photographic results) are now gradually unfolding themselves to astronomers. To put the matter in the fewest possible words there seems little or no doubt that according as spots on the Sun are abundant or scarce so the Corona when visible during an eclipse varies in appearance from one period of eleven years to another like period. Or, to put it in another way, given the date of a coming total eclipse we can predict to a certain extent the probable shape and character of the Corona if we know how the forthcoming date stands as regards a Sun-spot maximum or minimum.

The most recent important eclipses up to date which have been observed, namely those of April 16, 1893, Aug. 9, 1896, and Jan. 21, 1898, do not add much to our useful records of the outward appearances presented by the Corona. The 1896 Corona is described as intermediate between the two Types respectively associated with years of maximum and minimum Sun-spots, and this is as it should have been, albeit there was one extension which reached to about two diameters of the Sun. The 1898 Corona yielded four long Coronal streamers reaching much farther from the Sun than any previously seen, the two longest reaching to 4?and 6 diameters of the Sun respectively. These dimensions are quite unprecedented.

The application of the spectroscope to observations of eclipses of the Sun demands a few words of notice in this place, but it would not be consistent with the plan of this work to go into details. Though the spectroscope has been applied under many different circumstances to different parts of the Sun's surroundings in connection with total eclipses yet it is in regard to the Corona that most has been done and most has been discovered. The substance of the discoveries made is that the Corona shines with an intrinsic light of its own, that is to say, that it is composed of constituents whose temperature is sufficiently elevated to be self-luminous. These constituents are chiefly hydrogen; the body which corresponds to the line D3 (of Fraunhofer's scale), and which has been named "Helium"; and the body which corresponds to the bright green line 1474 of Kirchoff's scale and which, since its existence was first suspected and then assured, has been named "Coronium."

The reader will not be surprised to learn, from what has gone before, that an immense mass of records have accumulated respecting the appearance of the Corona. Correspondingly numerous and divergent are the theories which have been launched to explain the observations made. One thing is in the highest degree probable, namely, that electricity is largely concerned.

Going back to the question of Sun-spots regarded in their possible or probable association with the Corona, the present position of matters appears to be this: that there is a real connection between the general form of the Corona and disturbances on the Sun, taking Sun-spots as an indication of solar activity. When Sun-spots are at or near their maximum, the Corona has generally been somewhat symmetrical, with synclinal groups of rays making angles of 45?with its general axis. On the other hand, at the epochs of minimum Sun-spots, the Corona shows polar rifts much more widely open, with synclinal zones making larger angles with the axis, and being, therefore, more depressed towards the equatorial regions, in which, moreover, there is usually a very marked extension of Coronal matter in the form of elongated streamers reaching to several diameters of the Sun.

This generalisation is well borne out by the maximum-epoch Coronas of 1870 and 1871, and the minimum-epoch Coronas of 1867, 1874, 1875, 1878, and perhaps 1887, and certainly 1889. On the other hand, the eclipses of 1883, 1885 and 1886 do not strikingly confirm this theory. The eclipse of 1883 was at a time of rapidly decreasing solar activity, yet the Corona had the features of a Sun-spot maximum. The same, though in a somewhat less degree, may be said of the eclipses of 1885 and 1886. At the times of both of these eclipses the solar activity was decreasing.

The forthcoming eclipse of 1900 will nearly coincide with a Sun-spot minimum, and if the above conclusions are well founded the Corona in 1900 should resemble that of 1889, and be characterised by, amongst other things, some very elongated groups of rays extending in nearly opposite directions.

We are still a long way off from being able to state with perfect confidence what the Corona is. It is certainly a complex phenomenon, and the various streamers which we see are not, as was at one time imagined, a simple manifestation of one radiant light. Mrs. Todd thus conveniently summarises the present state of our knowledge:--"The true corona appears to be a triple phenomenon. First, there are the polar rays, nearly straight throughout their visible extent. Gradually, as these rays start out from points on the solar disc farther and farther removed from the poles, they acquire increasing curvature, and very probably extend into the equatorial regions, but are with great difficulty traceable there, because projected upon and confused with the filaments having their origin remote from the poles. Then there is the inner equatorial corona, apparently connected intimately with truly solar phenomena, quite like the polar rays; while the third element in the composite is the outer equatorial corona, made up of the long ecliptic streamers, for the most part visible only to the naked eye, also existing as a solar appendage, and possibly merging into the zodiacal light. The total eclipses of a half century have cleared up a few obscurities, and added many perplexities. There is little or no doubt about the substantial, if not entire, reality of the corona as a truly solar phenomenon. The Moon, if it has anything at all to do with the corona, aside from the fact of its coming in conveniently between Sun and Earth, so as to allow a brief glimpse of something startlingly beautiful which otherwise could never have been known, is probably responsible for only a very narrow ring of the inner radiance of pretty even breadth all round. This diffraction effect is accepted; but the problem still remains how wide this annulus may be, and whether it may vary in width from one eclipse to another. These questions once settled, the spurious structure may then be excerpted from the true. Indeed the coronal streamers, delicately curving and interlacing, may tell the whole story of the Sun's radiant energy."

FOOTNOTES:

[Footnote 16: There seems sufficient evidence to show that the Corona may be seen even on occasions when the Sun is not totally eclipsed, provided that the visible crescent of the Sun is exceedingly narrow.]

[Footnote 17: See p. 130 (post).]

CHAPTER VII.

WHAT IS OBSERVED AFTER THE TOTAL PHASE OF AN ECLIPSE OF THE SUN IS AT AN END.

In a certain sense, a description of the incidents which precede the total disappearance of the Sun in connection with a total Eclipse will apply more or less to the second half of the phenomenon; only, of course, in the reverse order and on the opposite side of the compass. The Corona having appeared first of all on the W. side of the Sun, then having shown itself complete as surrounding the Sun, will begin to disappear on the W. side, and will be last seen on the E. side. Baily's Beads may or may not come into view; the Sun will reappear first as a very thin crescent, gradually widening; the quasi-nocturnal darkness visible on the Earth will cease, and eventually the Moon will completely pass away from off the Sun, and the Sun once again will exhibit a perfect circle of light.

Whilst there is so much to look for and look at and think about, one thing must be sought for instantly after totality, or it will be gone for ever, and that is the Moon's shadow on the Earth. We have already seen in the last chapter the startling rapidity and solemnity with which the shadow seems to rush forward to the observer from the horizon on the western side of the Meridian. Passing over him, or even, so to speak, through him, it travels onwards in an easterly direction and very soon vanishes. Its visibility at all depends a good deal upon whether the observer, who is looking for it, is sufficiently raised above the adjacent country to be able to command at least a mile or two of ground. If he is in a hollow, he will have but little chance of seeing the shadow at all: on the other hand, if he is on the top of a considerable hill (or high up on the side of a hill), commanding the horizon for a distance of 10 or 20 miles, he will have a fair chance of seeing the shadow. Sir G. Airy states, in 1851, "My eye was caught by a duskiness in the S.E., and I immediately perceived that it was the Eclipse-shadow in the air, travelling away in the direction of the shadow's path. For at least six seconds, this shadow remained in sight, far more conspicuous to the eye than I had anticipated. I was once caught in a very violent hail and thunder-storm on the Table-land of the County of Sutherland called the "Moin," and I at length saw the storm travel away over the North Sea; and this view of the receding Eclipse-shadow, though by no means so dark, reminded me strongly of the receding storm. In ten or twelve seconds all appearance of the shadow had passed away."

Perhaps this may be a convenient place to make a note of what seems to be a fact, partly established at any rate, even if not wholly established, namely-- that there seems some connection between eclipses of the Sun and Earthquakes. A German physicist named Ginzel[18] has found a score of coincidences between solar eclipses and earthquakes in California in the years between 1850 and 1888 inclusive. Of course there were eclipses without earthquakes and earthquakes without eclipses, but twenty coincidences in thirty-eight years seems suggestive of something.

FOOTNOTES:

[Footnote 18: Himmel und Erde, vol. ii. pp. 255, 309; 1890.]

CHAPTER VIII.

ECLIPSES OF THE SUN MENTIONED IN HISTORY--CHINESE.

This is the first of several chapters which will be devoted to historical eclipses. Of course the total eclipse of the Sun of August 9, 1896, observed in Norway and elsewhere, is, in a certain sense, an eclipse mentioned in history, but that is not what is intended by the title prefixed to these chapters. By the term "historical eclipses," as used here, I mean eclipses which have been recorded by ancient historians and chroniclers who were not necessarily astronomers, and who wrote before the invention of the telescope. The date of this may be conveniently taken as a dividing line, so that I shall deal chiefly with eclipses which occurred before, say, the year 1600. There is another reason why some such date as this is a suitable one from which to take a new departure. Without at all avowing that superstition ceased on the Earth in the year 1600 (for there is far too large a residuum still available now, 300 years later), it may yet be said that the Revival of Letters did do a good deal to divest celestial phenomena of those alarming and panic-causing attributes which undoubtedly attached to them during the earlier ages of the world and during the "Dark Ages" in Western Europe quite as much as during any other period of the world's history. No one can examine the writings of the ancient Greek and Roman historians, and the chronicles kept in the monasteries of Western Europe by their monkish occupiers, without being struck by the influence of terror which such events as eclipses of the Sun and Moon and such celestial visitors as Comets and Shooting Stars exercised far and wide. And this influence overspread, not only the unlettered lower orders, but many of those in far higher stations of life who, one might have hoped, would have been exempt from such feelings of mental distress as they often exhibited. Illustrations of this fact will be adduced in due course.

It has always been supposed that the earliest recorded eclipse of the Sun is one thus mentioned in an ancient Chinese classic--the Chou-King (sometimes spelt Shou-Ching). The actual words used may be translated:--"On the first day of the last month of Autumn the Sun and Moon did not meet harmoniously in Fang." To say the least of it, this is a moderately ambiguous announcement, and Chinese scholars, both astronomers and non-astronomers, have spent a good deal of time in examining the various eclipses which might be thought to be represented by the inharmonious meeting of the Sun and the Moon as above recorded. To cut a long story short, it is generally agreed that we are here considering one or other of two eclipses of the Sun which occurred in the years 2136 or 2128 B.C. respectively, the Sun being then in the sidereal division "Fang," a locality determined by the stars [Greek: beta], [Greek: delta], [Greek: pi], and [Greek: rho]Scorpii, and which includes a few small stars in Libra and Ophiuchus to the N. and in Lupus to the S. How this simple and neat conclusion, which I have stated with such apparent dogmatism, was arrived at is quite another question, and it would hardly be consistent with the purpose of this volume to attempt to work it out in detail, but a few points presented in a summary form may be interesting.

In the first place, be it understood, that though it is fashionable to cast ridicule on John Chinaman, especially by way of retaliation for his calling us "Barbarians," yet it is a sure and certain fact that not only have the Chinese during many centuries been very attentive students of Astronomy, but that we Westerns owe a good deal of our present knowledge in certain departments to the information stored up by Chinese observers during many centuries both before and after the Christian Era.

This, however, is a digression. The circumstances of this eclipse as regards its identification having been carefully examined by Mr. R.Rothman,[19] in 1839 were further reviewed by Professor S.Russell in a paper published in the proceedings of the Pekin Oriental Society.[20] The substance of the case is that in the reign of Chung-K'ang, the fourth Emperor of the Hsia Dynasty, there occurred an eclipse of the Sun, which is interesting not only for its antiquity, but also for the dread fate of the two Astronomers Royal of the period, who were taken by surprise at its occurrence, and were unprepared to perform the customary rites. These rites were the shooting of arrows and the beating of drums, gongs, etc., with the object of delivering the Sun from the monster which threatened to devour it. The two astronomers by virtue of their office should have superintended these rites. They were, however, drunk and incapable of performing their duties, so that great turmoil ensued, and it was considered that the land was exposed to the anger of the gods. By way of appeasing the gods, and of suitably punishing the two State officials for their neglect and personal misconduct, they were forthwith put to death, a punishment which may be said to have been somewhat excessive, in view of the fact that the eclipse was not a total but only a partial one. An anonymous verse runs:--

Here lie the bodies of Ho and Hi, Whose fate though sad was visible-- Being hanged because they could not spy Th' eclipse which was invisible.

It appears beyond all reasonable doubt that the eclipse in question occurred on October 22, 2136 B.C. The preliminary difficulties to be got over in arriving at the date arose from the fact that there was an uncertainty of 108 years in the date when the Emperor Chung-K'ang ascended the throne; and within these limits of time there were 14 possible years in which an eclipse of the Sun in Fang could have occurred. Then the number was further limited by the necessity of finding an eclipse which could have been seen at the place which was the Emperor's capital. The site of this, again, was a matter of some uncertainty. However, step by step, by a judicious process of exhaustion, the year 2136 B.C. was arrived at as the alternative to the previously received date of 2128 B.C. Considering that we are dealing with a matter which happened full 4000 years ago, it may fairly be said that this discrepancy is not perhaps much to be wondered at, seeing what disputes often happen nowadays as to the precise date of events which may have occurred but a few years or even a few months before the controversy springs up.

Professor Russell says that:--"Some admirers of the Chinese cite this eclipse as a proof of the early proficiency attained by the Chinese in astronomical calculations. I find no ground for that belief in the text. Indeed, for many centuries later, the Chinese were unable to predict the position of the Sun accurately among the stars. They relied wholly on observation to settle their calendar, year by year, and seem to have drawn no conclusions or deductions from their observations. Their calendar was continually falling into confusion. Even at the beginning of this dynasty, when the Jesuits came to China, the Chinese astronomers were unable to calculate accurately the length of the shadow of the Sun at the equinoxes and solstices. It seems to me therefore very improbable that they could have been able to calculate and predict eclipses."

I am not at all sure that this is quite a fair presentation of the case. I do not remember ever to have seen the power to predict eclipses ascribed to the Chinese, but it is a simple matter of fact that we owe to them during many centuries unique records of a vast number of celestial phenomena. Their observations of comets may be singled out as having been of inestimable value to various 19th-century computers, especially E. Biot and J. Hind.

The second recorded eclipse of the Sun would seem to be also due to the Chinese. Confucius relates that during the reign of the Emperor Yew-Wang an eclipse took place. This Emperor reigned between 781 B.C. and 771 B.C., and it has been generally thought that the eclipse of 775 B.C. is the one referred to, but Johnson doubts this on the ground that this eclipse was chiefly visible in the circumpolar regions, and if seen at all in China must have been of very small dimensions. He leans to the eclipse of June 4, 780 B.C. as the only large one which happened within the limits of time stated above.

An ancient Chinese historical work, known as the Chun-Tsew, written by Confucius, makes mention of a large number of solar eclipses which occurred before the Christian Era. This work came under the notice of M. Gaubil, one of the French Jesuit missionaries who laboured in China some century and a half ago, and he first gave an account of it in his Trait?de la Chronologie Chinoise, published at Paris in 1770.[21]

The Chun-Tsew is said to be the only work really written by Kung-Foo-Tze, commonly known as Confucius, the other treatises attributed to him having been compiled by disciples of his either during his life-time or after his decease. The German chronologist, Ideler, was acquainted with this work, and in a paper of his own, presented to the Berlin Academy, remarked:--

"What gives great interest to this work is the account of 36 solar eclipses observed in China, the first of which was on Feb. 22, 720 B.C., and the last on July 22, 495 B.C."

In 1863 Mr. John Williams, then Assistant Secretary of the Royal Astronomical Society, communicated to the Society in a condensed form the particulars of these eclipses as related in Confucius's book, together with some remarks on the book itself. The Chun-Tsew treats of a part of the history of the confederated nations into which China was divided during the Chow Dynasty, that is between 1122 B.C. and 255 B.C. The particular period dealt with is that which extended from 722 B.C. to 479 B.C. It was during the latter part of this interval of about 242 years that Confucius flourished. But the book is not quite a general history for it is more particularly devoted to the small State of Loo of which Confucius was a native, where he passed a great portion of his life, and where he was advanced to the highest honours. It contains the history of twelve princes of this State with incidental notices of the other confederated nations. The number of the years of each reign is accurately determined, and the events are classed under the years in which they occurred. Each year is divided into sections according to the four seasons, Spring, Summer, Autumn, Winter, and the sections are subdivided into months, and often the days are distinguished. The name Chun-Tsew is said to have been given to this work from its having been commenced in Spring and finished in Autumn, but Williams thinks that the name rather refers to the fact that its contents are divided into seasons as stated. The style in which it is written is very concise, being a bare mention of facts without comment, and although on this account it might appear to us dry and uninteresting, it is much valued by the Chinese as a model of the ancient style of writing. It forms one of the Woo-King or Five Classical Books, without a thorough knowledge of which, and of the Sze-Shoo or Four Books, no man can attain to any post of importance in the Chinese Empire.

The account of each eclipse is but little more than a brief mention of its occurrence at a certain time. The following is an example of the entries:--"In the 58th year of the 32nd cycle in the 51st year of the Emperor King-Wang, of the Chow Dynasty, the 3rd year of Yin-Kung, Prince of Loo, in the spring, the second moon, on the day called Kea-Tsze, there was an eclipse of the Sun." This 58th year of the 32nd cycle answers to 720 B.C. Mr. Williams in the year 1863 presented to the Royal Astronomical Society a paper setting out the whole of the eclipses of which I have cited but one example, converting, of course, the very complicated Chinese dates into European dates.

These Chinese records of eclipses were in 1864 subjected to examination by the late Sir G. Airy,[22] with results which were highly noteworthy, and justify us in reposing much confidence in Chinese astronomical work. Airy remarks:-- "The period through which these eclipses extend is included in the time through which calculations of eclipses have been made in the French work entitled L'Art de vie fier les Dates. I have several times had occasion to recalculate with great accuracy eclipses which are noted in that work (edition of 1820), and I have found that, to the limits of accuracy to which it pretends, and which are abundantly sufficient for the present purpose, it is perfectly trustworthy. I have therefore made a comparison of the Chun-Tsew eclipses with those of L'Art de vie fier les Dates. The result is interesting. Of the 36 eclipses, 32 agree with those of the Art de vie fier les Dates, not only in the day, but also in the general track of the eclipse as given in the Art de vie fier, which appears to show sufficiently that the eclipse would be visible in that province of China to which the Chun-Tsew is referred." Airy then proceeds to point out that, with regard to the four eclipses which he could not confirm, there cannot have been eclipses in April 645 B.C. or in June 592 B.C. It appears, however, from a note by Williams, that the date attached to the eclipse of 645 B.C. is, in reality, an erroneous repetition (in the Chinese mode of expressing it) of that attached to the next following one, and in the absence of correct date it must be rejected. In the record of 592 B.C., June 16, no clerical error is found, and there must be an error of a different class. The eclipses of 552 B.C., September 19, and 549 B.C., July 18, to which there is nothing corresponding in the Art de vie fier, are in a different category. These occur in the lunations immediately succeeding 552 B.C., August 20, and 549 B.C., June 19, respectively, and there is no doubt that those which agree with the Art de vie fier were real eclipses. Now there cannot be eclipses visible at the same place in successive lunations, because the difference of the Moon's longitudes is about 29? and the difference of latitudes is therefore nearly 3? which is greater than the sum of the diameters of the Sun and Moon increased by any possible change of parallax for the same place. These, therefore, were not real eclipses. It seems probable that the nominal days were set down by the observer in his memorandum book as days on which eclipses were to be looked for. Airy conjectured that the eclipses of 552 B.C., August 20, and 549 B.C., June 19, were observed by one and the same person, and that he possessed science enough to make him connect the solar eclipses with the change of the Moon, but not enough to give him any idea of the limitations to the visibility of an eclipse.

On a subsequent occasion Mr. Williams laid before the Society a further list of solar eclipses observed in China, and extending from 481 B.C. to the Christian Era. He collected these from a Chinese historical work, entitled Tung-Keen-Kang-Muh. This work, which runs to 101 volumes, contains a summary of Chinese history from the earliest times to the end of the Yuen Dynasty, A.D. 1368, and was first published about 1473. The copy in Mr. Williams's possession was published in 1808. The text is very briefly worded, and consists merely of an account of the accessions and deaths of the emperors and of the rulers of the minor states, with some of the more remarkable occurrences in each reign. The appointments and deaths of various eminent personages are also noticed, together with special calamities such as earthquakes, inundations, storms, etc. The astronomical allusions include eclipses and comets. Amongst the eclipses are also all, or most of those which are recorded in the Chun-Tsew as having occurred prior to 479 B.C. Though no particular expressions are used to define the exact character of the eclipses, it is to be presumed that some of them must have been total, because it is stated that the stars were visible, albeit that seemingly in only one instance is a word attached which specifically expresses the idea of totality. Here again all the dates were expressed in Chinese style, but, as published by Williams, were rendered, as before, in European style by aid of chronological tables, published about 1860 in Japan. Mr. Williams, in his second paper, from which I have been quoting, states that he brought his published account down to the Christian Era only as a matter of convenience, but that he had in hand a further selection of eclipses from the Tung-Keen-Kang-Muh, the interval from the Christian Era to the 4th century A.D. yielding nearly 100 additional eclipses. This further transcript has not yet been published, but remains in MS. in the Library of the Royal Astronomical Society. Mr. Williams died in 1874 at the age of 77, one of the most experienced Chinese scholars of the century.

It is remarkable that none of the Chinese annals to which reference has been made include any mention of eclipses of the Moon; but the records of Comets are exceedingly numerous and, as I have already stated, have proved of the highest value to astronomers who have been called upon to investigate the ancient history of Comets.

FOOTNOTES:

[Footnote 19: Memoirs, R.A.S., vol. xi. p. 47.]

[Footnote 20: Republished in the Observatory Magazine, vol. xviii. p. 323, et seq., 1895.]

[Footnote 21: A good deal of information respecting Chinese eclipse records, so far as known up to the beginning of the 19th-century, will be found in Delambre's Histoire de l'Astronomie Ancienne. Paris, 1817.]

[Footnote 22: Month. Not., R.A.S., vol. xxiv. p. 41.]

CHAPTER IX.

ARE ECLIPSES ALLUDED TO IN THE BIBLE?

An interesting question has been suggested: Are there any allusions to eclipses to be found in Holy Scripture? It seems safe to assert that there is at least one, and that there may be three or four.

In Amos viii. 9 we read:--"I will cause the Sun to go down at noon, and I will darken the Earth in the clear day." This language is so very explicit and applies so precisely to the circumstances of a solar eclipse that commentators are generally agreed that it can have but one meaning;[23] and accordingly it is considered to refer without doubt to one or other of the following eclipses:-- 791 B.C., 771 B.C., 770 B.C., or 763 B.C. Archbishop Usher,[24] the well-known chronologist, suggested the first three more than two centuries ago, whilst the eclipse of 763 B.C. was suggested in recent times and is now generally accepted as the one referred to. The circumstances connected with the discovery and identification of the eclipse of 763 B.C. are very interesting.

The date when Amos wrote is set down in the margin of our Bibles as 787 B.C. and if this date is correct it follows that for his statement to have been a prediction he must be alluding to some eclipse of later date than 787 B.C. This obvious assumption not only shuts out the eclipse of 791 B.C., but opens the door to the acceptance of the eclipse of 763 B.C.

Apparently the first modern writer who looked into the matter after Archbishop Usher was the German commentator Hitzig who suggested the eclipse of Feb. 9, 784 B.C. Dr. Pusey was so far taken with this idea that he thought it worth while to secure the co-operation of the Rev. R. Main, F.R.A.S., the Radcliffe Observer at Oxford, for the purpose of a full investigation. Mr. Main had the circumstances of that eclipse calculated, with the result that though the eclipse was indeed total in Africa and Hindostan, yet at Samaria it was only partial and of no considerable magnitude. Dr. Pusey's words, summing up the situation are:--"The eclipse then would hardly have been noticeable at Samaria, certainly very far indeed from being an eclipse of such magnitude, as could in any degree correspond with the expression, 'I will cause the Sun to go down at noon.'" ... "Beforehand, one should not have expected that an eclipse of the Sun, being itself a regular natural phenomenon, and having no connection with the moral government of God, should have been the subject of the prophet's prediction. Still it had a religious impressiveness then, above what it has now, on account of that wide-prevailing idolatry of the Sun. It exhibited the object of their false worship, shorn of its light, and passive."

Dr. Pusey's Commentary from which the above quotation is made[25] bears the date 1873, but he appears not to have been acquainted with the important discovery announced no less than six years previously by the distinguished Oriental scholar, Sir H. Rawlinson. The discovery to which I allude is a contemporary record on an Assyrian tablet of a solar eclipse which was seen at Nineveh about 24 years after the reputed date of Amos's prophecy. This tablet had been described by Dr. Hinckes in the British Museum Report for 1854 but its chronological importance had not then been realised. Sir H. Rawlinson[26] speaks of the tablet as a record of or register of the annual archons at Nineveh. He says:--"In the eighteenth year before the accession of Tiglath-Pileser there is a notice to the following effect--'In the month Sivan an eclipse of the Sun took place' and to mark the great importance of the event a line is drawn across the tablet although no interruption takes place in the official order of the Eponymes. Here then we have notice of a solar eclipse which was visible at Nineveh which occurred within 90 days of the (vernal) equinox (taking that as the normal commencement of the year) and which we may presume to have been total from the prominence given to the record, and these are conditions which during a century before and after the era of Nabonassar are alone fulfilled by the eclipse which took place on June 15, 763."

This record was submitted to Sir G. Airy and Mr. J. Hind, and the circumstances of the eclipse were computed by the latter, by the aid of Hansen's Lunar Tables and Le Verrier's Solar Tables. The result, when plotted on a map, showed that the shadow line just missed the site of Nineveh, but that a very slight and unimportant deviation from the result of the Tables would bring the shadow over the city of Nineveh where the eclipse was observed, and over Samaria where it was predicted. The identification of this eclipse, both as regards its time and place, has also proved a matter of importance in the revision of Scripture chronology, by lowering, to the extent of 25 years, the reigns of the kings of the Jewish monarchy. The need for this revision is further confirmed, if we assume that the celebrated incident in the life of King Hezekiah, described as the retrogradation of the Sun's shadow on the dial of Ahaz, is to be interpreted as connected with a partial eclipse of the Sun.

We will now consider this event, and see what can be made out of it. One Scripture record (2 Kings xx. 11) is as follows:--"And Isaiah the prophet cried unto the Lord: and he brought the shadow ten degrees backward, by which it had gone down in the dial of Ahaz." This passage has greatly exercised commentators of all creeds in different ages of the Church; and the most divergent opinions have been expressed as to what happened. This has been due to two causes jointly. Not only is the occurrence incomprehensible, looked at on the surface of the words, but we are entirely ignorant of the construction of the so-called "dial" of Ahaz, and have little or no material directly available from outside sources to enable us to come to a clear and safe conclusion. No doubt, however, it was a sun-dial, or gnomon of some kind. Bishop Wordsworth lays stress on the apparent assertion that the miracle was not wrought on any other dial at Jerusalem except that of Ahaz, the father of Hezekiah, and he treats as a confirmation of this the statement in 2 Chron. xxxii. 31, that ambassadors came from Babylon to Jerusalem, being curious to learn all about "the wonder that had been done in the land" (i.e. in the land of Judah). But there is more taken for granted here than is necessary, or, as we shall presently see, is justifiable. To begin with, how do we know that there was any other dial at Jerusalem like that of Ahaz? But, in point of fact, we must make a new departure altogether, for it has been suggested (I know not exactly by whom, or when for the first time) that an eclipse of the Sun, under certain circumstances, would explain all that happened, and reconcile all that has to be reconciled. What happened to Hezekiah is thought by many to imply clearly a miracle, and it may be said that an eclipse of the Sun cannot be held to be a miracle[27] by the ordinary definition of the word. But, on the other hand, it certainly might count as such in the eyes of ignorant spectators, who know nothing of the theory or practice of eclipses, and who would regard such a thing as quite unforeseen, unexpected, and alarming. Illustrations of this might be multiplied from all parts of the world, in all ages of the world's history.

Let us see now what the argument is, as it was worked out by the late Mr. J. Bosanquet, F.R.A.S. Shortly before the invasion of Judea by Sennacherib--say in the beginning of the year 689 B.C.--Hezekiah was sick unto death. In answer to his fervent prayer for recovery the prophet Isaiah was sent to him with this message:--"Thus saith the Lord, the God of David thy Father, I have heard thy prayer, I have seen thy tears; behold, I will add unto thy days fifteen years ... and I will defend this city, and this shall be a sign unto thee from the Lord, that the Lord will do this thing that He hath spoken. Behold, I will bring again the shadow of the degrees, which is gone down in the sun-dial of Ahaz ten degrees backward. So the Sun returned ten degrees, by which degrees it had gone down." (Isaiah xxxviii. 5-8).

In these words we evidently have mention of some instrument erected in Hezekiah's palace, in the days of his father Ahaz, for showing the change in the position of the shadow cast by the Sun from day to day. This statement is confirmed by a profane writer, Glycas, who states: "They say that Ahaz, by some contrivance, had erected in his palace certain steps, which showed the hours of the day, and also measured the course of the Sun."

The idea involved in "bringing again," through "ten degrees backward," "the shadow of the degrees" which had gone down, is very noteworthy. We seem intended to learn from these words several things. For one thing (to begin with) that the steps (as we must consider them to have been) on this sun-dial of Ahaz, were turned away from the Sun. For only in that position could they cast their shadow, or could the number of the illuminated steps be varied, upwards or downwards, according to the varying altitude of the sun. The only conceivable use of a fixed instrument so placed would be to show the rise and fall of the shadow from day to day, as the Sun on the meridian gradually rose higher between mid-winter and mid-summer, or descended lower between mid-summer and mid-winter, in passing of course through the winter and summer solstices in turn. No simple motion of the Sun in its ordinary diurnal progress would produce the effect described. On the other hand, it is equally clear that the shadow cast by a gnomon properly adjusted at the head of such a series of steps would travel upwards and downwards upon the steps "with the Sun," from winter to summer and from summer to winter, indicating at each noon the meridian altitude of the Sun from day to day, the latitude of Jerusalem being 31?47', and the Sun's altitude there on the shortest day being 34degree41'. If the gnomon were raised above the topmost step so as to bring the tip of the gnomon or any aperture in it so much above the step as would be the equivalent of 2degree54' or slightly more, then the top of the shadow of the gnomon (or a spot of light passing through a hole in it) would, on the shortest day of the year, fall just beyond the lowermost step. An instrument constructed on the principle just set forth was known to and used by the Greek astronomers of antiquity under the name of a Sciotheron or shadow-taker. Sometimes, and perhaps more properly, it was called a Heliotropion, that is, an instrument designed to indicate the turning of the Sun at the Tropics.[28] This, be it remembered, was information needed by the ancients for the correct regulation of the seasons of the year, and of special service to the Jews whose greater festivals were fixed in connection with the seasons. There is reason to believe that instruments of this character were of early invention, going back perhaps to the times of Homer, for we find a passage in the Odyssey, (xv. 403) as follows:--

"Above Ortygia lies an isle of fame Far hence remote, and Syria [Syros] is the name; There curious eyes inscrib'd with wonder trace The Sun's diurnal and his summer race."

Pope's rendering of this passage fails, however, to bring out the salient idea involved. Butcher and Lang translate the passage thus:--"There is a certain isle called Syria, if haply thou hast heard tell of it, over above Ortygia, and there are the turning-places of the Sun." Merry[29] calls these island names mere "inventions of the poet." It seems to me a great question whether Homer's words really support the statement I have made just before quoting it.

Diogenes Latius refers to this same instrument when he speaks of the Heliotropion preserved in the Island of Syra.[30]

According to Latius, Anaximander[31] was the first Greek to use gnomons, which he placed on the Sciothera of Laced 鎚 on, for the express purpose of indicating the Tropics and Equinoxes. These Sciothera were pyramidal in form.

An obelisk was the simplest, though an imperfect form of Heliotropion, marking indistinctly the length of a shadow at different times of the year, especially the extremes of length and shortness at mid-winter and mid-summer. It is perhaps interesting to mention that travellers have recorded, in various places, various devices for furnishing information respecting these matters. For instance, in Milan Cathedral the meridian line is marked on the pavement, and along this line, an image of the Sun coming through an aperture in the southern wall travels backwards and forwards during the year according to the seasons. Some Jesuit missionaries who visited China about the middle of the last century, noticed a device of this character in operation at the Observatory at Pekin. A gnomon had been set up in a low room and one of the missionaries, M. Le Comte, describes in the following words what they saw in connection with this gnomon:--"The aperture through which the rays of the Sun came was about 8 ft. above the floor; it is horizontal and formed of two pieces of copper, which may be turned so as to be farther from, or closer to, each other to enlarge or contract the aperture. Lower was a table with a brass plate in the middle on which was traced a meridian line 15 ft. long, divided by transverse lines which are neither finished nor exact. All round the table there are small channels to receive the water, whereby it is to be levelled."[32]

All this may seem rather a digression, and so it is, but I am following Mr. Bosanquet herein in order the better to justify the argument that it was an eclipse of the Sun which marked the important incident in Hezekiah's life which has been handed down to us by the sacred writer. It is evident that if a flight of steps were erected on the principles which were set forth above, the steps sloping upwards and southwards (for the Northern Hemisphere) from the lowest step to within a few inches below an aperture in the gnomon suitably arranged, the ray or image of the Sun, whichever it was, would travel day by day up and down such steps between solstice and solstice. We may conclude, therefore, that the instrument which Hezekiah gazed at, and which is called in Scripture, the "Dial" of Ahaz, was what the Greeks would have termed a Heliotropion.

The historian's record is to the effect that on the day of Hezekiah's recovery an extraordinary motion of the shadow was observed on the "Steps of Ahaz" by the rising of the shadow "ten steps" from the point to which it had "gone down with the Sun." This effect is spoken of not as a miracle but as "a sign." It should also be remembered that the cure of Hezekiah was effected not by a miracle but by a simple application of a lump of figs. The promise of his recovery was confirmed by the motion of the shadow as already stated. We are justified, therefore, in looking for some ordinary natural phenomenon by which to account for this peculiar motion on the dial, and something miraculous is not essential. Dean Milman once suggested that the effect might have been produced "by a cloud refracting the light." No doubt a dark cloud might produce an apparent interference with the shadow, but it is well pointed out by Bosanquet that such a cause as a cloud would have been so manifest to everyone, and the effect so transient, that the phenomenon could hardly have been referred to afterwards as it was in another place as "a wonder that was done in the land." (2 Chron. xxxii. 31).

It becomes, therefore, alike an obvious and a simple explanation that a shadow caused by the Sun might be deflected downwards on such an instrument with a regular and steady motion by the Moon passing slowly over the upper part of the Sun's disc, as Sun and Moon both approached the meridian.

The critical question has now to be raised: "Can astronomers inform us whether a considerable eclipse of the Sun occurred at the beginning of the year 689 B.C. anywhere near noon and which was visible at Jerusalem?" And the answer to this it is interesting to be able to say is a plain and distinct affirmative. There was a large partial eclipse of the Sun on January 11, 689 B.C., about 11.30 A.M., and it was the upper limb which underwent eclipse.

This eclipse fulfils all the requirements of the case, both from the historian's and the astronomer's point of view. It occurred about the year fixed by Demetrius as that of Hezekiah's illness: it occurred while the Sun was approaching and actually passing the meridian; the obscuration was on that part of the Sun's disc (namely the upper part) which would have had the effect of causing the point of light, which would seem to emanate from the Sun, to appear to be depressed downwards; and it was visible at Jerusalem. But there still remains for consideration the final and most important question, "Would a deflection of light proceeding from the Sun, regarded as a moving body, be capable of affecting, to the extent of 'ten steps,' the shadow on such an instrument as has been described?" And arising out of this, there is the subordinate question, "Would January, being the month when this eclipse certainly occurred, also be a month suitable for the exhibition of such a phenomenon?"

It is ascertainable by calculation that the time occupied by the Moon in passing over the Sun, in the way it did during this eclipse, was about 2?hours. But from the time of central conjunction, when the obscuration was the greatest and the point of light depressed the most, to the time when the uppermost portion of the Sun's disc was released by the eastward motion of the Moon, and the light from that uppermost portion was again manifest, was about 20 minutes, and this, therefore, was the time during which the phenomenon of retrogression on the "steps" would have been exhibited to the King's eyes. Assuming then that the time when the ascending shadow had travelled upwards to the tenth step coincided, or nearly so, with the time when the Sun had reached its highest altitude for the day, at noon, we infer that the time of central conjunction during this eclipse was not later than from 20 to 15 minutes before noon. It could not have been much earlier, because the phenomenon of the resting of the shadow for a time at its apparently highest point for the day (which preceded the promise that it should rise ten steps) has also to be accounted for, and this cessation of its motion upwards could not have taken place till about 25 minutes before noon, when the decreasing motion of the Sun in altitude (or its slackening motion upwards as it approached mid-day) would have become counteracted by the coming on of the eclipse. Now at 11.35 A.M. the sun's disc would have risen to the altitude of 35degree8'; and the highest visible point of light would, owing to the eclipse, then have been about 35degree4'; and at 11.40 A.M., being the time of greatest obscuration, the extreme cusps of light produced by the intervention of the Moon would still have stood at about 35degree4', just 23' below the highest point of light at noon (Fig. 12). The whole disc of the sun had now risen above the gnomon, yet no motion of the shadow on the steps had been observed for fully five minutes. The time shown by the dial was seemingly mid-day.

Sun's apparent semi-diameter 16' 13" Moon's " " 15' 13" Moon's relative hourly motion in declination 5' 44" northward. Right ascension, 29' 33" eastward. Corrected for Jerusalem, 19' 42" eastward. Altitude of the Gnomon, $34\degree41'$ 13".

SUN'S ALTITUDE BEFORE AND AT NOON.

We have now to consider "to what extent would a staircase rising at an angle of $31\degree47'$ towards the Sun, with a gnomon so placed at the top as to cast a shadow to the foot of the lower step on the shortest day of the year be affected by a movement in a perpendicular direction of the point of light to the extent of 23', or 1/3 of a degree"? The effect would be widely different at different times of the year, being greatest at mid-winter when the shadows are longest, and least at mid-summer when the shadows are shortest. It follows from this that January 13 being a day but three weeks removed from mid-winter day the normal shadow would be not far from its longest possible length, and the effect of a displacement of 23' would be neither more nor less than 1/12th of the whole range of the steps whatever that range might have been. This extent of motion, then, is fully sufficient to satisfy the condition prescribed by the Biblical narrative of there being such a deflection of the Sun's light as would affect the shadow to the extent implied by the words "ten steps" or "ten degrees," which is virtually the same idea. The same extent of motion could not have been produced under the same conditions either a few days earlier or a few days later; that may certainly be taken for granted. And the only point in which we are necessarily in doubt arises from the fact that we are ignorant of the actual number and nature of the graduations of Ahaz's so-called "Dial." If it were permissible to assume that there were 120 graduations on the instrument, be they steps properly so-called on a structure erected in the open air or be they lines on a flat surface on some instrument standing in a room, or what not, then the problem is solved, for 1/12 (as above) of 120 is ten--the "ten degrees" stated in the history.

As to whether the "dial" of Ahaz was a device built up of masonry in the open air or was an instrument for indoor use we know absolutely nothing, and speculation is useless. There is something to be said on both sides. Bosanquet, on abstract grounds, leans to the latter view; on the other hand he calls attention to the present existence in India, at Delhi and Benares, of ruined Hindoo observatories in the form of huge masonry sun-dials many yards in length and breadth and height.[33]

Finally it may be pointed out that there is some incidental confirmation to be found for this Hezekiah incident having happened in winter. That the season of the year was winter seems to be suggested by the word used in the original Hebrew in connection with the return of the shadow.

"Backward" in Isaiah xxxviii. 8 might also be translated, "From the end." It would be very natural to hold that this implied that the motion of the shadow was upwards from the lower end of the group of steps towards which the shadow had gone down. Now the lower end of the steps could only have been the place of the shadow in December or January at or near the time of the winter solstice. Moreover the mention of the "lump of figs" seems to suggest the winter season. A cake of figs means dried figs, not newly gathered summer figs.

Putting all the facts together we may fairly conclude that the astronomical event which happened in connection with Hezekiah's illness was an eclipse of the Sun, and that its date was January 11, 689 B.C.

A few other Scripture passages need a passing mention. In Isaiah xiii. 10 we read:--

"The Sun shall be darkened in his going forth, and the Moon shall not cause her light to shine." It has been thought by Johnson that this passage is an allusion to an eclipse of the Sun, and so it might be; but on the other hand, it may be no more than one of those highly figurative phrases which abound in holy Scripture, and of which the well-known passage, "The stars in their courses fought against Sisera" (Judges v. 20), is a familiar example.

In Jeremiah x. 2 we read:--

"Be not dismayed at the signs of heaven; for the heathen are dismayed at them." This is cited as an eclipse allusion by Johnson, who points out that the utterance of this caution preceded by about fifteen years the celebrated eclipse of Thales (585 B.C.). But surely this is far-fetched. I shall be inclined to attach the same criticism to his next citation. Ezekiel employs these expressions:--"When I shall put thee out, I will cover the heaven, and make the stars thereof dark; I will cover the Sun with a cloud, and the Moon shall not give her light" (xxxii. 7). This language resembles, in no small degree, Isaiah's, already quoted, and, like that, might apply to the phenomenon of a solar eclipse, but whether that was actually the prophet's intention is another matter. He may have witnessed the eclipse of 585 B.C. on the banks of the river Chebar, and that spectacle may have put this imagery into his head. Further than this it seems hardly safe to go.

This seems an appropriate place to mention a very interesting matter, to which attention has been called by Oriental scholars in recent times, who have investigated Assyrian and Egyptian monuments, and other monuments of the same type. The story would be a long and interesting one if presented in detail, and would far exceed my limits of space. I must, therefore, be content with such a summary as that which has been worked out by Mr. E. Maunder. Briefly the facts are these. There are to be found in many places carvings in stone, symbolic of the Sun-god once worshipped in the East. The general design, with of course variations, is a circle with striated wings extending right and left to two diameters of the wing, more or less, with a lesser extension in a downward direction. Allowing for the roughness of the art, and for the fact that the material was stone, it does not require any very great stretch of imagination to see in these carvings the disc of a totally-eclipsed Sun with, right and left and below it, that form of corona which we have come to associate with total eclipses occurring at periods of Sun-spot minima.[34] This idea should not seem far-fetched if we bear in mind the fact that the ancient Orientals worshipped the Sun, Moon, and Planets; and one of the natural outcomes of this is submitted for our consideration by Maunder in the words following[35]:--

"There can be little doubt that the Sun was regarded partly as a symbol, partly as a manifestation of the unseen, unapproachable Divinity. Its light and heat, its power of calling into active exercise the mysterious forces of germination and ripening, the universality of its influence, all seemed the fit expressions of the yet greater powers which belonged to the Invisible. What happened in a total solar eclipse? For a short time that which seemed so perfect a divine symbol was completely hidden. The light and heat, the two great forms of solar energy, were withdrawn, but something took their place. A mysterious light of mysterious form, unlike any other light, unlike any other single form, was seen in its place. Could they fail to see in this a closer, a more intimate revelation, a more exalted symbolism of the Divine Nature and Presence? Just as in the various Greek 'mysteries' the student was gradually advanced from one set of symbols to another even more abstruse and esoteric, so here, on the broad face of heaven itself, vouchsafed for a brief space of time and at long intervals apart, the Deity revealed Himself to the initiated by a higher and more difficult symbol than ordinarily. The symbol would vary in shape. We may take it for granted that the old Chaldeans, as modern astronomers to-day, had at one time or another presented to them every type of Coronal structure. But there would, no doubt, be a difficulty in grasping or remembering the irregular details of the Corona as seen in most eclipses. It occasionally happens, however, that the Corona shows itself under a form of grand and striking simplicity. It is now widely recognised that the typical Corona of the minimum of the Sun-spot cycle consists chiefly of two great equatorial streamers."

Maunder then goes on to cite certain American pictures by Trouvelot and others of the eclipse of July 29, 1878, in which the great extension of the Corona to the East and the West is specially shown. One drawing in particular, by Miss K.Wolcott, exhibits the Sun with a perfect bright ring round it from which the Coronal streamers emanate in the directions mentioned. Maunder then remarks that he has a strong conviction that it was a Corona of this type which was the origin of the "Ring with Wings," the symbol which on Assyrian monuments is always shown as floating over the head of the ring which is designed to indicate the presence and protection of the Deity. In the article cited he gives illustrations of two forms under which the "Ring with Wings" appears on Assyrian and Egyptian monuments respectively, remarking that "Egyptians too were Astronomers and Sun-worshippers and were experts in the language of symbols. Equally with the Chaldeans the Egyptian priests should have regarded the Corona as a symbolical revelation of the Deity whose usual manifestation they recognised in the Sun, and accordingly we find them employing a symbol which is almost as perfect a representation of the Corona of minimum as that which the Assyrians adopted." Another curious point commented upon by Maunder is that the Assyrians frequently insert the figure of their Deity within the ring, and attach thereto a kilt-like dress. Even when they show the ring without the figure the "kilt," as it may be called, is still there, indicating that it is not simply a garment worn by the figure, but an integral part of the symbol. This "kilt" is represented as pleated, and the resemblance of the pleatings to the polar rays shown in Trouvelot's drawing of the Corona, is "practically perfect." On this point Maunder adds:-- "If this be a mere chance coincidence, it seems to me a most extraordinary one." He concludes by saying that these symbols, so frequently met with, and so clearly designed to indicate the presence of the Deity, "are, in their origin, drawings of the solar Corona, as seen at the Sun-spot minimum, and as such are the earliest eclipse representations which have been preserved to us."

I give these ideas for what they are worth; they are very ingeniously worked out, and though the argument is not conclusive, yet I do think that there is enough in it to be worth attention.

FOOTNOTES:

[Footnote 23: Less certain is the allusion in Amos v. 8:--"Seek him that ... maketh the day dark with night."]

[Footnote 24: Annales, A.M., 3213, p. 45. Folio Ed.]

[Footnote 25: Minor Prophets, p. 217.]

[Footnote 26: Athenem, May 18, 1867.]

[Footnote 27: After all, do the circumstances necessarily presuppose a "miracle"? Hezekiah had only asked for a "sign." In 2 Chron. xxxii. 31 the word "wonder" is applied to the event.]

[Footnote 28: Hence the word "Tropic," from [Greek: trep (I turn).]

[Footnote 29: Homer, Odyssey, vol. ii. p. 255. Clarendon Press Series.]

[Footnote 30: Life of Pherecydes, sec. 6.]

[Footnote 31: Life of Anaximander, sec. 3.]

[Footnote 32: Du Halde's "China," 3rd edition, 1741, vol. iii. p. 86.]

[Footnote 33: Paper by W. Hunter in Asiatic Researches, vol. v., p. 190. The Benares Observatory is described by Sir R. Barker in Phil. Trans., vol. lxvii., p. 598. 1777.]

[Footnote 34: See p. 70 (ante).]

[Footnote 35: Knowledge, vol. xx., p. 9, January 1897.]

CHAPTER X.

ECLIPSES OF THE SUN MENTIONED IN HISTORY--CLASSICAL.

In this chapter we shall, for the most part, be on firmer ground than hitherto, because several of the most eminent Greek and Latin historians have left on record full and circumstantial accounts of eclipses which have come under their notice, and which have been more or less completely verified by the computations and researches of astronomers in modern times. But these remarks do not, however, quite apply to the first eclipse which will be mentioned.

Plutarch, in his Life of Romulus, refers to some remarkable incident connected, in point of time at any rate, with his death:--"The air on that occasion was suddenly convulsed and altered in a wonderful manner, for the light of the Sun failed, and they were involved in an astonishing darkness, attended on every side with dreadful thunderings and tempestuous winds." This so-called darkness is considered to have been the same as that mentioned by Cicero.[36] There is so much myth about Romulus that it is not safe to write in confident language. Nevertheless it is a fact, according to Johnson, that there was a very large eclipse of the Sun visible at Rome in the afternoon of May 26, 715 B.C., and 715 B.C. is supposed to have been the year, or about the year, of the death of Romulus. Plutarch is also responsible for the statement that a great eclipse of the Sun took place sometime before the birth of Romulus; and if there is anything in this statement Johnson thinks that the annular eclipse of November 28, 771 B.C., might meet the circumstances of the case, but too much romance attaches to the history of Romulus for anyone to write with assurance respecting the circumstances of his career. Much of it is generally considered to be fabulous.

In one of the extant fragments of the Greek poet Archilochus (said to be the first who introduced iambics into his verses), the following sentence occurs:-- "Zeus the father of the Olympic Gods turned mid-day into night hiding the light of the dazzling sun; an overwhelming dread fell upon men." The poet's language may evidently apply to a total eclipse of the Sun; and investigations by Oppolzer and Millosevich make it probable that the reference is to the total eclipse of the Sun which happened on April 6, 648 B.C. This was total at about 10 a.m. at Thasos and in the northern part of the Aegean Sea. The acceptance of this date displaces by about half a century the date commonly assigned for the poet's career, but this is not thought to be of much account having regard to the hazy character of Grecian chronology before the Persian wars.[37]

On May 28, 585 B.C. there occurred an eclipse the surrounding circumstances of which present several features of particular interest. One of the most celebrated of the astronomers of antiquity was Thales of Miletus, and his astronomical labours were said to have included a prediction of this eclipse, which moreover has the further interest to us that it has assisted chronologists and historians in fixing the precise date of an important event in ancient history. Herodotus[38] describing a war which had been going on for some years between the Lydians and the Medes gives the following account of the circumstances which led to its premature termination:--"As the balance had not inclined in favour of either nation, another engagement took place in the sixth year of the war, in the course of which, just as the battle was growing warm, day was suddenly turned into night. This event had been foretold to the Ionians by Thales of Miletus, who predicted for it the very year in which it actually took place. When the Lydians and Medes observed the change they ceased fighting, and were alike anxious to conclude peace." Peace was accordingly agreed upon and cemented by a twofold marriage. "For (says the historian) without some strong bond, there is little security to be found in men's covenants." The exact date of this eclipse was long a matter of discussion, and eclipses which occurred in 610 B.C. and 593 B.C. were each thought at one time or another to have been the one referred to. The question was finally settled by the late Sir G. Airy, after an exhaustive inquiry, in favour of the eclipse of 585 B.C. This date has the further advantage of harmonising certain statements made by Cicero and Pliny as to its having happened in the 4th year of the 48th Olympiad.

Another word or two may be interesting as regards the share which Thales is supposed to have had in predicting this eclipse, the more so, that very high authorities in the domains of astronomy, and chronology, and antiquities take opposite sides in the matter. Sir Lewis, Bart., M.P., may be cited first as one of the unbelievers. He says[39] that Thales is "reported to have predicted it to the Ionians. If he had predicted it to the Lydians, in whose country the eclipse was to be total, his conduct would be intelligible, but it seems strange that he should have predicted it to the Ionians who had no direct interest in the event." Bosanquet replies to this by pointing out that Miletus, in Ionia, was the birthplace of Thales, and also that a shadow, covering two degrees of latitude, passing through Ionia, would also necessarily cover Lydia.

Another dissentient is Sir H.Rawlinson,[40] who, remembering that Thales is said to have predicted a good olive crop, and Anaxagoras says:--"The prediction of this eclipse by Thales may fairly be classed with the prediction of a good olive crop. Thales, indeed, could only have obtained the requisite knowledge for predicting eclipses from the Chaldeans; and that the science of these astronomers, although sufficient for the investigation of lunar eclipses, did not enable them to calculate solar eclipses--dependent as such a calculation is, not only on the determination of the period of recurrence, but on the true projection also of the track of the Sun's shadow along a particular line over the surface of the earth--may be inferred from our finding that in the astronomical canon of Ptolemy, which was compiled from the Chaldean registers, the observations of the Moon's eclipse are alone entered."

Airy[41] replied to these observations as follows:--"I think it not at all improbable that the eclipse was so predicted, and there is one easy way, and only one of predicting it--namely, by the Saros, or period of 18 years, 10 days, 8 hours nearly. By use of this period an evening eclipse may be predicted from a morning eclipse but a morning eclipse can rarely be predicted from an evening eclipse (as the interval of eight hours after an evening eclipse will generally throw the eclipse at the end of the Saros into the hours of night). The evening eclipse, therefore, of B.C. 585, May 28, which I adopt as being most certainly the eclipse of Thales, might be predicted from the morning eclipse of B.C. 603, May 17.... No other of the eclipses discussed by Baily and Oltmanns present the same facility for prediction."

Xenophon[42] mentions an eclipse as having led to the capture by the Persians of the Median city Larissa. In the retreat of the Greeks on the eastern side of the Tigris, they crossed the river Zapetes and also a ravine, and then reached the Tigris. According to Xenophon, they found at this place a large deserted city formerly inhabited by the Medes. Its wall was 25 feet thick and 100 feet high; its circumference 2 parasangs [=??miles]. It was built of burnt brick on an under structure of stone 20 feet in height. Xenophon then proceeds to say that "when the Persians obtained the Empire from the Medes, the King of the Persians besieged the city but was unable by any means to take it till a cloud having covered the Sun and caused it to disappear completely, the inhabitants withdrew in alarm, and thus the city was captured. Close to this city was a pyramid of stone, one plethrum in breadth, two plethra in height.... Thence the Greeks proceeded six parasangs to a great deserted castle by a city called Mespila formerly inhabited by the Medes; the substructure of its wall was of squared stone abounding in shells ... the King of the Persians besieged it but could not take it; Zeus terrified the inhabitants with thunderbolts, and so the city was taken."

The minute description here given by Xenophon enabled Sir A. Layard, Captain Felix Jones, and others, to identify Larissa with the modern Nimrud and Mespila with Mosul. A suspicion is thrown out in some editions of the Anabasis that the language cited might refer to an eclipse of the Sun. It is to be noted, however, that it is not included by Ricciolus in the list of eclipses mentioned in ancient writers which he gives in his Almagestum Novum. Sir G. Airy, having had his attention called to the matter, examined roughly all the eclipses which occurred during a period of 40 years, covering the supposed date implied by Xenophon. Having selected two, he computed them accurately but found them inapplicable. He then tried another (May 19, 557 B.C.) which he had previously passed over because he doubted its totality, and he had the great satisfaction of finding that the eclipse, though giving a small shadow, had been total, and that it had passed so near to Nimrud that there could be no doubt of its being the eclipse sought.

Sir G. Airy was such a very careful worker and investigator of eclipses that his conclusions in this matter have met with general acceptance. It must, however, in fairness be stated that a very competent American astronomer, Professor Newcomb, has expressed doubts as to whether after all

Xenophon's allusion is to an eclipse, but, judging by his closing words, the learned American does not seem quite satisfied with his own scepticism, for he says--"Notwithstanding my want of confidence, I conceive the possibility of a real eclipse to be greater than in the eclipse of Thales, while we have the great advantages that the point of occurrence is well defined, the shadow narrow, and, if it was an eclipse at all, the circumstance of totality placed beyond serious doubt."[43]

In the same year as that in which, according to the common account, the battle of Salamis was fought (480 B.C.), there occurred a phenomenon which is thus adverted to by Herodotus[44]--"At the first approach of Spring the army quitted Sardis and marched towards Abydos; at the moment of its departure the Sun suddenly quitted its place in the heavens and disappeared though there were no clouds in sight and the day was quite clear; day was thus turned into night." We are told[45] that "As the king was going against Greece, and had come into the region of the Hellespont, there happened an eclipse of the Sun in the East; this sign portended to him his defeat, for the Sun was eclipsed in the region of its rising, and Xerxes was also marching from that quarter." So far as words go these accounts admirably befit a total eclipse of the Sun, but regarded as such it has given great trouble to chronologers, and the identification of the eclipse is still uncertain. Hind's theory is that the allusion is to an eclipse and in particular to the eclipse of February 17, 478 B.C. Though not total at Sardis yet the eclipse was very large, 94/100ths of the Sun being covered. If we accept this, it follows that the usually recognised date for the battle of Salamis must be altered by two years. Airy thought it "extremely probable" that the narrative related to the total eclipse of the Moon, which happened on March 13, 479 B.C., but this is difficult to accept, especially as Plutarch, in his Life of Pelopidas, says--"An army was soon got ready, but as the general was on the point of marching, the Sun began to be eclipsed, and the city was covered with darkness in the daytime." This seems explicit enough, assuming the record to be true and that the same incident is referred to by Plutarch as by Herodotus and Aristides.

Since the time when Airy and Hind examined this question, all the known facts have been again reviewed by Mr. W. Lynn, who pronounces, but with some hesitation, in favour of the eclipse of October 2, 480 B.C., as the one associated with the battle of Salamis. He does this by refusing to see in the above quotations from Herodotus any allusion to a solar eclipse at all, but invites us to consider a later statement in Herodotus[46] as relating to an eclipse though the historian only calls it a prodigy.

After the battle of Thermopyl?the Peloponnesian Greeks commenced to fortify the isthmus of Corinth with the view of defending it with their small army against the invading host of Xerxes. The Spartan troops were under the command of Cleombrotus, the brother of Leonidas, the hero of Thermopyl? He had been consulting the oracles at Sparta, and Herodotus states that "while he was offering sacrifice to know if he should march out against the Persian, the Sun was suddenly darkened in mid-sky." This occurrence so frightened Cleombrotus that he drew off his forces and returned home. It is uncertain from the narrative of Herodotus whether Cleombrotus returned to Sparta in the autumn of the year of the battle of Salamis, or in the spring of the next following year which was that in which the battle was fought. Bishop Thirlwall[47] thinks that it was the latter, but Lynn pronounces for the former, adding that the date may well have been in October, and the solar eclipse of October 2, 480 B.C. may have been the phenomenon which attracted notice, particularly as the Sun would have been high in the heavens, the greatest phase (6/10ths) occurring, according to Hind, at 50 minutes past noon. Here I must leave the matter, merely remarking that this alternative explanation obviates the necessity for disturbing the commonly received date of the battle of Salamis.

Thucydides states that during the Peloponnesian war "things formerly repeated on hearsay, but very rarely confirmed by facts, became not incredible, both about earthquakes and eclipses of the Sun which came to pass more frequently than had been remembered in former times." One such eclipse he assigns to the first year of the war and says[48] that "in the same summer, at the beginning of a new lunar month (at which time alone the phenomenon seems possible) the Sun was eclipsed after mid-day, and became full again after it had assumed a crescent form and after some of the stars had shone out." Aug. 3, 431 B.C. is generally recognised as the date of this event. The eclipse was not total only three-fourths of the Sun's disc being obscured. Venus was 20?and Jupiter 43?distant from the Sun, so probably these were the "stars" that were seen. This eclipse nearly prevented the Athenian expedition against the Lacedonians. The sailors were frightened by it, but a happy thought occurred to Pericles, the commander of the Athenian forces. Plutarch, in his Life of Pericles, says:--"The whole fleet was in readiness, and Pericles on board his own galley, when there happened an eclipse of the Sun. The sudden darkness was looked upon as an unfavourable omen, and threw the sailors into the greatest consternation. Pericles observing that the pilot was much astonished and perplexed, took his cloak, and having covered his eyes with it, asked him if he found anything terrible in that, or considered it as a bad presage? Upon his answering in the negative, he said, 'Where is the difference, then between this and the other, except that something bigger than my cloak causes the eclipse?'"

Another eclipse is mentioned by Thucydides[49] in connection with an expedition of the Athenians against Cythera. He says:--"At the very commencement of the following summer there was an eclipse of the Sun at the time of a new moon, and in the early part of the same month an earthquake." This has been identified with the annular eclipse of March 21, 424 B.C., the central line of which passed across Northern Europe. It is not quite clear whether the historian wishes to insinuate that the eclipse caused the earthquake or the earthquake the eclipse.

An eclipse known as that of Ennius is another of the eclipses antecedent to the Christian Era which has been the subject of full modern investigation, and the circumstances of which are such that, in the language of Professor Hansen, "it may be reckoned as one of the most certain and well-established eclipses of antiquity." The record of it has only been brought to light in modern times by the discovery of Cicero's Treatise, De Republic? According to Cicero,[50] Ennius the great Roman poet, who lived in the second century B.C., and who died of gout contracted, it is said, by frequent intoxication, recorded an interesting event in the following words:--Nonis Junii soli luna obstetit et nox, "On the Nones of June the Moon was in opposition to the Sun and night." This singular phrase has long been assumed to allude to an eclipse of the Sun, but the precise interpretation of the words was not for a long time realised. In Cicero's time the Nones of June fell on the 5th, but in the time of Ennius, who lived a century and a half before Cicero, the Nones of June fell between June 5 and July 4 on account of the lunar years and the intercalary month of the Roman Calendar. The date of this eclipse is distinctly known to be June 21, 400 B.C., but the hour was long in dispute. Professor Zech found that the Sun set at Rome eclipsed, and that the maximum phase took place after sun-set. Hansen, however, with his better Tables, found that the eclipse was total at Rome, and that the totality ended at 7.33 p.m., the Sun setting almost immediately afterwards at 7.36. This fact, Hansen considers, explains the otherwise unintelligible passage of Ennius quoted above: instead of saying et nox, he should have said et simul nox, "and immediately it was night." Newcomb questions the totality of this eclipse, but assigns no clear reasons for his doubts.[51]

On August 14, 394 B.C., there was a large eclipse of the Sun visible in the Mediterranean. It occurred in the forenoon, and is mentioned by Xenophon[52] in connection with a naval engagement in which the Persians were defeated by Conon.

Plutarch, in his Life of Pelopidas, relates how one, Alexander of Pher? had devastated several cities of Thessaly, and that as soon as the oppressed inhabitants had learned that Pelopidas had come back from an embassy on which he had been to the King of Persia, they sent deputies to him to Thebes to beg the favour of armed assistance, with Pelopidas as general. "The Thebans willingly granted their request, and an army was soon got ready, but as the general was on the point of marching, the Sun began to be eclipsed, and the city was covered with darkness in the day-time." This eclipse is generally identified with that of July 13, 364 B.C. If this is correct, Plutarch's language must be incorrect, or at least greatly exaggerated, for no more than about three-fourths of the Sun was obscured.

On February 29, 357 B.C., there happened an eclipse, also visible in or near the Mediterranean. This is supposed to have been the eclipse for the prediction of which Helicon, a friend of Plato, received from Dionysius, King of Syracuse, payment in the shape of a talent.

We have now to consider another ancient eclipse which has a history of peculiar interest as regards the investigations to which it has been subjected. It is commonly known as the "Eclipse of Agathocles," and is recorded by two historians of antiquity in the words following. Diodorus Siculus[53] says:--

"Agathocles also, though closely pursued by the enemy, by the advantage of the night coming on (beyond all hope), got safe off from them. The next day there was such an eclipse of the Sun, that the stars appeared everywhere in the firmament, and the day was turned into night, upon which Agathocles's soldiers (conceiving that God thereby did foretell their destruction) fell into great perplexities and discontents concerning what was like to befall them."

 Justin says[54]:--

"By the harangue the hearts of the soldiers were somewhat elevated, but an eclipse of the Sun that had happened during their voyage still possessed them with superstitious fears of a bad omen. The king was at no less pain to satisfy them about this affair than about the war, and therefore he told them that he should have thought this sign an ill presage for them, if it had happened before they set out, but having happened afterwards he could not but think it presaged ill to those against whom they marched. Besides, eclipses of the luminaries always signify a change of affairs, and therefore some change was certainly signified, either to Carthage, which was in such a flourishing condition, or to them whose affairs were in a very ruinous state."

 The substance of these statements is that in the year 310 B.C. Agathocles, Tyrant of Syracuse, while conducting his fleet from Syracuse to the Coast of Africa, found himself enveloped in the shadow of an eclipse, which evidently, from the accounts, was total. His fleet had been chased by the Carthaginians on leaving Syracuse the preceding day, but got away under the cover of night. On the following morning about 8 or 9 a.m. a sudden darkness came on which greatly alarmed the sailors. So considerable was the darkness, that numerous stars appeared. It is not at the first easy to localise the position of the fleet, except that we may infer that it could hardly have got more than 80 or at the most 100 miles away from the harbour of Syracuse where it had been closely blockaded by a Carthaginian fleet. Agathocles would not have got away at all but for the fact that a relieving fleet was expected, and the Carthaginians were obliged to relax their blockade in order to go in search of the relieving fleet. Thus it came about not only that Agathocles set himself free, but was able to retaliate on his enemies by landing on the coast of Africa at a point near the modern Cape Bon, and devastating the Carthaginian territories. The voyage thither occupied six days, and the eclipse occurred on the second day. Though we are not informed of the route followed by Agathocles, that is to say whether he passed round the North or the South side of the island of Sicily, yet it has been made clear by astronomers that the southern side was that taken.

Baily, who was the first modern astronomer to investigate the circumstances of this eclipse, found that there was an irreconcilable difference between the path of the shadow found by himself and the historical statement, a gap of about 180 geographical miles seeming to intervene between the most southerly position which could be assigned to the fleet of Agathocles, and the most northerly possible limit of the path of the eclipse shadow. This was the condition of the problem when Sir G.Airy took it up in 1853.[55] He, however, was able to throw an entirely new light upon the matter. The tables used by Baily were distinctly inferior to those now in use, and Sir G. Airy thought himself justified in saying that to obviate the discordance of 180 miles just referred to "it is only necessary to suppose an error of 3' in the computed distances of the Sun and Moon at conjunction, a very inconsiderable correction for a date anterior to the epoch of the tables by more than twenty-one centuries."

It deserves to be mentioned, though the point cannot here be dwelt upon at much length, that these ancient eclipses all hang together in such a way that it is not sufficient for the man of Astronomy and the man of Chronology to agree on one eclipse, unless they can harmonise the facts of several.

For instance, the eclipse of Thales, the date of which was long and much disputed, has a material bearing on the eclipse of Agathocles, the date of which admits of no dispute; and one of the problems which had to be solved half a century ago was how best to use the eclipse of Agathocles to determine the date of that of Thales. If 610 B.C. were accepted for the Thales eclipse, so as to throw the zone of total darkness anywhere over Asia Minor (where for the sake of history it was essential to put it) the consequence would be that the shadow of the eclipse of 310 B.C. would have been thrown so far on to land, in Africa, as to make it out of the question for Agathocles and his fleet to have been in it, yet we know for a certainty that he was in it in that year, and no other year. Conversely, if 603 B.C. were accepted for the Thales eclipse, then to raise northwards the position of the shadow in that year from the line of the Red Sea and the Persian Gulf, that it might pass through Asia Minor, would so raise the position of the shadow in 310 B.C. as to throw it far too much to the N. of Sicily for Agathocles, who we know must have gone southwards to Africa, to have entered it. But if we assume 585 B.C. as the date of the eclipse of Thales, we obtain a perfect reconciliation of everything that needs to be reconciled; the shadow of the eclipse of 585 B.C.

will be found to have passed where ancient history tells us it did pass--namely, through Ionia, and therefore through the centre of Asia Minor, and on the direct route from Lydia to Media; whilst we also find that the shadow of the 310 B.C. eclipse, that is the one in the time of Agathocles, passed within 100 miles of Syracuse, a fact which is stated almost in those very words by the two historians who have recorded the doings of Agathocles and his fleet in those years.

This is where the matter was left by Airy in 1853. Four years later the new solar and lunar tables of the German astronomer Hansen were published, and having been applied to the eclipse of 585 B.C., the conclusions just stated were amply confirmed. As if to make assurance doubly sure, Airy went over his ground again, testing his former conclusions with regard to the eclipse of Thales by the eclipse of Larissa, in 557 B.C. already referred to, and bringing in the eclipse of Stiklastad in 1030 A.D., to be referred to presently. And as the final result, it may be stated that all the foregoing dates are now known to an absolute certainty, especially confirmed as they were in all essential points by a computer of the eminence of the late Mr. J. Hind.

On a date which corresponds to February 11, 218 or 217 B.C., an eclipse of the Sun, which was partial in Italy, is mentioned by Livy.[56] Newcomb found that the central line passed a long way from Italy, to wit, "far down in Africa."

An eclipse of the Sun is mentioned by Dion Cassius[57] as having happened when Caesar crossed the Rubicon, a celebrated event made use of by speakers, political and otherwise, on endless occasions in modern history. There seems no doubt that the passage of the Rubicon took place in 51 B.C., and that the eclipse must have been that of March 7, 51 B.C. The circumstances of this eclipse have been investigated by Hind, who found that the eclipse was an annular one, the annular phase lasting 6?minutes in Northern Italy.

Arago associates the death of Julius Caesar in 44 B.C. with an annular eclipse of the Sun, but seemingly without sufficient warrant. The actual record is to the effect that about the time of the great warrior's death there was an extraordinary dimness of the Sun. Whatever it was that was noticed, clearly it could not have been an annular eclipse, because no such eclipse then happened. Johnson suggests that Arago confused the record of some meteorological interference with the Sun's light with the annular eclipse that happened seven years previously when Caesar passed the Rubicon, to which eclipse allusion has already been made. That there was for a long while a great deficiency of sunshine in Italy about the time of Caesar's death seems clear from remarks made by Pliny, Plutarch, and Tibullus, and the words of Suetonius seem to imply something of a meteorological character. I should not have mentioned this matter at all, but for Arago's high repute as an astronomer. According to Seneca[58] during an eclipse a comet was also seen.

It is an interesting question to inquire whether any allusions to eclipses are to be found in Homer, and no very certain answer can be given. In the Iliad (book xvii., lines 366-8) the following passage will be found:--"Nor would you say that the Sun was safe, or the Moon, for they were wrapt in dark haze in the course of the combat."

In the Odyssey (book xx., lines 356-7) we find:--"And the Sun has utterly perished from heaven and an evil gloom is overspread." This was considered by old commentators to be an allusion to an eclipse, and in the opinion of W.Merry[59] "this is not impossible, as they were celebrating the Festival of the New Moon."

Certainly this language has somewhat the savour of a total eclipse of the Sun, but it is difficult to say whether the allusion is historic, as of a fact that had happened, or only a vague generality. Perhaps the latter is the most justifiable surmise.

I have in the many preceding pages been citing ancient eclipses, for the reason, more or less plainly expressed, that they are of value to astronomers as assisting to define the theory of the Moon's motions in its orbit, and this they should do; but it is not unreasonable to bring this chapter to a close by giving the views of an eminent American astronomer as to the objections to placing too much reliance on ancient accounts of eclipses. Says Prof. S. Newcomb[60]:--"The first difficulty is to be reasonably sure that a total eclipse was really the phenomenon observed. Many of the statements supposed to refer to total eclipses are so vague that they may be referred to other less rare phenomena. It must never be forgotten that we are dealing with an age when accurate observations and descriptions of natural phenomena were unknown, and when mankind was subject to be imposed upon by imaginary wonders and prodigies. The circumstance which we should regard as most unequivocally marking a total eclipse is the visibility of the stars during the darkness. But even this can scarcely be regarded as conclusive, because Venus may be seen when there is no eclipse, and may be quite conspicuous in an annular or a considerable partial eclipse. The exaggeration of a single object into a plural is in general very easy. Another difficulty is to be sure of the locality where the eclipse was total. It is commonly assumed that the description necessarily refers to something seen where the writer flourished, or where he locates his story. It seems to me that this cannot be safely done unless the statement is made in connection with some battle or military movement, in which case we may presume the phenomena to have been seen by the army."

FOOTNOTES:

[Footnote 36: De Republic? Lib. vi., cap. 22.]

[Footnote 37: E. Millosevich, Memorie della Societa Spettroscopisti Italiani, vol. xxii. p. 70. 1893.]

[Footnote 38: Herodotus, Book i., chap. 74. This eclipse is also mentioned by Pliny (Nat. Hist., Book ii., chap. 9) and by Cicero (De Divinatione, cap. 49).]

[Footnote 39: Astronomy of the Ancients, p. 88.]

[Footnote 40: Herodotus, edited by Rev. G. Rawlinson, vol. i. p. 212.]

[Footnote 41: Month. Not., R.A.S., vol. xviii. p. 148; March 1858.]

[Footnote 42: Anabasis, Lib. iii., cap. 4, sec. 7.]

[Footnote 43: Washington Observations, 1875, Appendix II., p. 31.]

[Footnote 44: Book vii., chap. 37. See Rawlinson's Herodotus, vol. iv. p. 39.]

[Footnote 45: Scholia, in Aristidis Orationes, Ed. Frommel, p. 222.]

[Footnote 46: Book ix., chap. 10. See Rawlinson's Herodotus, 3rd ed. vol. iv.

p. 379.]

[Footnote 47: History of Greece, vol. ii. p. 330.]

[Footnote 48: Book ii., chap. 28.]

[Footnote 49: Book iv., chap. 52.]

[Footnote 50: De Republic? Lib. i. c. 16.]

[Footnote 51: Washington Observations, 1875, Appendix II., p. 33.]

[Footnote 52: Hellenics, Book iv., chap. 3, sec. 10.]

[Footnote 53: Bibliothec?Historic? Lib. xx., cap. 1, sec. 5.]

[Footnote 54: Historia, Lib. xxii., cap. 6.]

[Footnote 55: Phil. Trans., vol. cxliii. pp. 187-91, 1853.]

[Footnote 56: Hist. Rom., Lib. xxii., cap. 1.]

[Footnote 57: Hist. Rome, Book xli., chap. 14.]

[Footnote 58: Naturalium Questionum, Lib. vii.]

[Footnote 59: Homer, Odyssey, vol. ii. p. 328. Clarendon Press Series.]

[Footnote 60: Washington Observations, 1875, Appendix II., p. 18.]

CHAPTER XI.

ECLIPSES OF THE SUN MENTIONED IN HISTORY.-- THE CHRISTIAN ERA TO THE NORMAN CONQUEST.

The Christian Era is, for several reasons, a suitable point of time from which to take a new departure in speaking of historical eclipses, although the First Century, at least, might obviously be regarded as belonging to classical history--but let that pass.

Dion Cassius[61] relates that on a date corresponding to March 28, A.D. 5, the Sun was partly eclipsed. Johnston says that the central line passed over Norway and Sweden. It seems, perhaps, a little strange that a writer who lived in Bithynia in the 3rd Century of the Christian Era should have picked up any information about something that happened in the extreme North of Europe two centuries previously. But probably the eclipse must have been seen in Italy.

On November 24, A.D. 29, there happened an eclipse of the Sun which is sometimes spoken of as the "eclipse of Phlegon." Eusebius, the ecclesiastical historian, records Phlegon's testimony. Phlegon was a native of Tralles in Lydia, and one of the Emperor Adrian's freedmen. The eclipse in question happened at noon, and the stars were seen. It was total, and the line of totality, according to Hind,[62] passed across the Black Sea from near Odessa to Sinope, thence near the site of Nineveh to the Persian Gulf. A partial eclipse with four-fifths of the Sun's diameter covered was visible at Jerusalem. This is the only solar eclipse which was visible at Jerusalem during the period usually fixed for Christ's public ministry. This eclipse was for a long time, and by various writers, associated with the darkness which prevailed at Jerusalem on the day of our Lord's Crucifixion, but there seems no warrant whatever for associating the two events. The Crucifixion darkness was assuredly a supernatural phenomenon, and there is nothing supernatural in a total eclipse of the Sun. To this it may be added that both Tertullian at the beginning of the 3rd century and Lucian, the martyr of Nicomedia, who died in 312, appealed to the testimony of national archives then in existence, as witnessing to the fact that a supernatural darkness had prevailed at the time of Christ's death. Moreover, the generally recorded date of the Crucifixion, namely, April 3, A.D. 33, would coincide with a full Moon. As it happened, that full Moon suffered eclipse, but she emerged from the Earth's shadow about a quarter of an hour before she rose at Jerusalem the penumbra continued upon her disc for an hour afterwards.

Speaking of the Emperor Claudius, Dion Cassius[63] says:--"There was going to be an eclipse on his birthday. Claudius feared some disturbance, as there had been other prodigies, so he put forth a public notice, not only that the obscuration would take place and about the time and magnitude of it, but also about the causes which produce such events." This is an interesting statement, especially in view of what I have said on a previous page about the indifference of the Romans to Astronomy. It would, likewise, be interesting to know how Claudius acquired his knowledge, and who coached him up in the matter. This eclipse occurred on August 1, A.D. 45. Barely half the Sun's diameter was covered.

Philostratus[64] states that "about this time while he was pursuing his studies in Greece such an omen was observable in the heavens. A crown resembling Iris surrounded the disc of the Sun and darkened its rays." "About this time" is to be understood as referring to some date shortly preceding the death of the Emperor Domitian which occurred on September 18, A.D. 96. This has usually been regarded as the earliest allusion to what we now call the Sun's "Corona"; or, as an alternative idea, that the allusion is simply to an annular eclipse of the Sun. But both these theories have been called in question; by Johnston because he cannot find an eclipse which in his view of things will respond as regards date to the statement of Philostratus, and by Lynn on the same ground and on other grounds, more suo. The question of identification requires looking into more fully. There was a total eclipse on May 21, A.D. 95, but it was only visible as a partial eclipse in Western Asia and not visible at all in Greece. This is given as the conclusion arrived at by the German astronomer Ginzel. But it does not seem to me sufficient to overthrow, without further investigation, the fairly plain language of Philostratus, which is possibly confirmed by a passage in Plutarch[65] in which he discusses certain eclipse phenomena in the light of a recent eclipse. The date of Plutarch's "recent" eclipse is somewhat uncertain, but that fact does not necessarily militate against his testimony respecting the Corona or what is regarded to have been such. The statement of Philostratus, treated as a mention of a total solar eclipse, is accepted as sufficiently conclusive by Sir W. Huggins and the late Professor R. Grant. Johnston, to meet the supposed difficulty of finding an eclipse to accord with the assertion of the historian, suggests that "perhaps some peculiar solar halo or mock Sun, or other meteorological formation" is referred to. But Stockwell has advanced very good reasons for the opinion that the eclipse of Sept. 3, A.D. 118, fully meets the circumstances of the case. Grant's opinion is given in these emphatic words:--"It appears to me that the words here quoted [from Apollonius] refer beyond all doubt to a total eclipse of the Sun, and thus the phenomenon seen encompassing the Sun's disc was, really as well as verbally, identical with the modern Corona."[66]

With the end of the first century of the Christian Era we may be said to quit the realms of classical history and to pass on to eclipse records of a different character, and, so far as regards European observations, of comparatively small scientific value or usefulness. Our information is largely derived from ecclesiastical historians and, later on, from monkish chronicles, which as a rule are meagre in a surprising degree. Perhaps I ought not to say "surprising," because after the times of the Greek astronomers (who in their way may almost be regarded as professionals), and after the epoch of the famous Ptolemy, Astronomy well-nigh ceased to exist for many centuries in Europe, until, say, the 15th century, barring the labours of the Arabians and their kinsmen the Moors in Spain in the 9th and following centuries.

In examining therefore the records of eclipses which have been handed down to us from A.D. 100 forwards through more than 1000 years, I shall not offer my readers a long dry statement of eclipse dates, but only pick out here and there such particular eclipses as seem to present details of interest for some or other reason.

On April 12, 237 A.D., there was, according to Julius Capitolinus, an eclipse of the Sun, so great "that people thought it was night, and nothing could be done without lights." Ricciolus remarked that this eclipse happened about the time of the Sixth Persecution of the Christians, and when the younger Gordian was proclaimed Emperor, after his father had declined the proffered dignity, being 80 years of age. The line of totality crossed Italy about 5 p.m. in the afternoon, to the N. of Rome, and embraced Bologna.

Calvisius records, on the authority of Cedrenus, an eclipse of the Sun on August 6, 324 A.D., which was sufficiently great for the stars to be seen at mid-day. The eclipse was associated with an earthquake, which shattered thirteen cities in Campania. Johnston remarks that no more than three-fourths of the Sun's disc would have been covered, as seen in Campania, but that elsewhere in Italy, at about 3 p.m., the eclipse was much larger, and perhaps one or two of the planets might have been visible.

On July 17, 334 A.D., there was an eclipse, which seems to have been total in Sicily, if we may judge from the description given by Julius Firmicus.[67]

Ammianus Marcellinus[68] describes an eclipse, to which the date of August 28, 360 A.D., has been assigned. Humboldt, quoting this historian, says that the description is quite that of a solar eclipse, but its stated long duration (daybreak to noon), and the word caligo (fog or mist) are awkward factors. Moreover, the historian associates it with events which happened in the eastern provinces of the Roman Empire; but Johnston seems in effect to challenge Marcellinus's statement when he says, "It is true that there was an annular eclipse of the Sun in the early morning on the above date, but it could only be seen in countries E. of the Persian Gulf."

About the time that Alaric, King of the Visigoths appeared before Rome, there was a gloom so great that the stars appeared in the daytime. This narrative is considered to apply to an eclipse of the Sun, which occurred on June 18, 410 A.D. The eclipse was an annular one, but as the central line must have crossed far S. of Rome, the stars must have been seen not at Rome but somewhere else.

An eclipse occurred on July 19, 418 A.D., which is remarkable for a twofold reason. People had an opportunity not only of seeing an eclipse, but also a comet. We owe the account of the circumstances to Philostorgius,[69] who tells us that--"On July 19, towards the 8th hour of the day, the Sun was so eclipsed, that even the stars were visible. But at the same time that the Sun was thus hid, a light, in the form of a cone was seen in the sky; some ignorant people called it a comet, but in this light we saw nothing that announced a comet, for it was not terminated by a tail; it resembled the flame of a torch, subsisting by itself, without any star for its base. Its movement too was very different from that of a comet. It was first seen to the E. of the equinoxes; after that, having passed through the last star in the Bear's tail, it continued slowly its journey towards the W. Having thus traversed the heavens, it at length disappeared, having lasted more than four months. It first appeared about the middle of the summer, and remained visible until nearly the end of autumn."

Boillot, a French writer, has suggested that this description is that of the zodiacal light, but this seems out of the question in view of the details given by the Chinese of a comet having been visible in the autumn of this year for 11 weeks, and having passed through the square of Ursa Major. Reverting to the eclipse--Johnston finds that the greatest phase at Constantinople, which was probably the place of observation, occurred at about half an hour after noon, when a thin crescent of light might have been seen on the northern limb of the Sun. From this it would appear that the central line of eclipse must have passed somewhat to the south of Constantinople. To the same effect Hind, who found that 95/100ths of the Sun's diameter was covered at Constantinople.

An eclipse of the Sun seems to be referred to by Gregorius Turonensis, when he says[70] that:--"Then even the Sun appeared hideous, so that scarcely a third part of it gave light, I believe on account of such deeds of wickedness and shedding of innocent blood." This would seem to have been the eclipse which occurred on February 24, 453 A.D., when Attila and the Huns were ravaging Italy, and to them it was doubtless that the writer alluded. At Rome three-fourths of the Sun's disc would have been eclipsed at sunset, a finding which tallies fairly with the statement of Gregorius.

It is not till far into the 6th century that we come upon a native English record of an eclipse of the Sun as having been observed in England. This deficiency in our national annals is thus judiciously explained and commented on by our clever and talented American authoress.[71] Speaking of the eclipse of February 15, 538 A.D., she says:--"The accounts, however, are greatly confused and uncertain, as would perhaps be natural fully 60 years before the advent of St. Augustine, and when Britain was helplessly harassed with its continual struggle in the fierce hands of West Saxons and East Saxons, of Picts and conquering Angles. Men have little time to record celestial happenings clearly, much less to indulge in scientific comment and theorising upon natural phenomena, when the history of a nation sways to and fro with the tide of battle, and what is gained to-day may be fatally lost to-morrow. And so there is little said about this eclipse, and that little is more vague and uncertain even than the monotonous plaints of Gildas--the one writer whom Britain has left us, in his meagre accounts of the conquest of Kent, and the forsaken walls and violated shrines of this early epoch."

The well-known Anglo-Saxon Chronicle[72] is our authority for this eclipse having been noted in England, but the record is bare indeed:--"In this year the Sun was eclipsed 14 days before the Calends of March from early morning till 9 a.m." Tycho Brahe, borrowing from Calvisius, who borrowed from somebody else, says that the eclipse happened "in the 5th year of Henry, King of the West Saxons, at the 1st hour of the day till nearly the 3rd, or immediately after sunrise." Johnson finds that at London nearly three-fourths of the Sun's disc was covered at 7.43 a.m.

The next eclipse recorded in the Anglo-Saxon Chronicle is somewhat difficult to explain. It is said that in 540 A.D. "The Sun was eclipsed on the 12th of the Calends of July and the stars appeared full nigh half an hour after 9 a.m." Johnson's calculations make the middle of the eclipse to have occurred at about 7.37 a.m. at London, two-thirds of the Sun's diameter being covered. He notes that the Moon's semi-diameter was nearly at its maximum whilst the Sun's semi-diameter was nearly at its minimum--a favourable combination for a long totality. The visibility of the stars seems difficult to explain in connection with this eclipse, and therefore he suggests that the annalist has made a mistake of four years and meant to refer to the eclipse of September 1, 536 A.D., but this does not seem a satisfactory theory.

The year after Pope Martin held a Synod to condemn the Monothelite heresy, an eclipse of the Sun took place. It is mentioned by Tycho Brahe in his catalogue of eclipses as having been seen in England. Johnson gives the date as February 6, 650 A.D., and finds that the Sun was three-fourths obscured at London at 3.30 p.m.

The Anglo-Saxon Chronicle tells us under the year A.D. 664 that, "In this year the Sun was eclipsed on the 5th of the Nones of May; and Earcenbryht, King of the Kentish people died and Ecgbryht his son succeeded to the Kingdom." Kepler thought this eclipse had been total in England, and Johnson calculating for London found that on May 1, at 5 p.m., there would only have been a very thin crescent of the Sun left uncovered on the southern limb, so that the line of totality would have passed across the country some distance to the N. of London.

The eclipse of Dec. 7, A.D. 671, seems to be associated with a comic tragedy. The Caliph Moawiyah had a fancy to remove Mahomet's pulpit from Medina to his own residence at Damascus. "He said that the walking-stick and pulpit of the Apostle of God should not remain in the hands of the murderers of Othman. Great search was made for the walking-stick, and at last they found it. Then they went in obedience to his commands to remove the pulpit, when immediately, to their great surprise and astonishment, the Sun was eclipsed to that degree that the stars appeared."[73] Once again the question of visible stars is in some sense a source of difficulty. Hind found that the eclipse was annular on the central line. At Medina the greatest phase occurred at 10h.?3m.m. when 85/100ths of the Sun's diameter was obscured. Hind suggests that in the clear skies of that part of the world such a degree of eclipse might be sufficient to bring out the brighter planets or stars. At any rate no larger eclipse visible at Medina occurred about this epoch. Prof. Ockley seems to refer to this eclipse in making, on the authority of several Arabian writers, the mention he does of an eclipse in the quotation just given.

Perhaps this will be a convenient place to bring in some remarks on certain Arabian observations of eclipses only made known to the scientific world in modern times. That the Arabians were very capable practical astronomers has long been recognised as a well-established fact, and if it had not been for them there would have been a tremendous blank in the history of astronomy during at least six centuries from about the year A.D. 700 onwards. In the year 1804 there was published at Paris a French translation of an Arabian manuscript preserved at the University of Leyden of which little was known until near the end of the last century. The manuscript was then sent to Paris on loan to the French Government which caused a translation to be made by "Citizen" Caussin, and this was published under the title of Le Livre de la grande Table Hakate.[74] Caussin was Professor of Arabic at the College of France. Newcomb considers this to contain the earliest exact astronomical observations of eclipses which have reached us. He remarks that some of the data left us by Ptolemy, Theon, Albategnius and others may be the results of actual observations, but in no case, so far as is known, have the figures of the actual observations been handed down. For example, we cannot regard "midnight" nor "the middle of an eclipse" as moments capable of direct observation without instruments of precision; but in the Arabian work under consideration we find definite statements of the altitudes of the heavenly bodies at the moments of the beginning and ending of eclipses--data not likely to be tampered with in order to agree with the results of calculation. The eclipses recorded are 28 in number and usually the beginning and end of them were observed. The altitudes are given sometimes only in whole degrees, sometimes in coarse fractions of a degree. The most serious source of error to be confronted in turning these observations to account arises from the uncertainty as to how long after the first contact the eclipse was perceived and the altitude taken; and how long before the true end was the eclipse lost sight of. Making the best use he could of the records available Newcomb found that they could safely be employed in his investigations into the theory of the Moon.

The observations were taken, some at Bagdad and the remainder at Cairo. I do not propose to occupy space by transcribing the accounts in detail, but one extract may be offered as a sample of the rest--"Eclipse of the Sun observed at Bagdad, August 18, 928 A.D. The Sun rose about one-fourth eclipsed. We looked at the Sun on a surface of water and saw it distinctly. At the end when we found no part of the Sun was any longer eclipsed, and that its disc appeared in the water as a complete circle, its altitude was 12?in the E., less the one-third of a division of the instrument, which itself was divided to thirds of a degree. One must therefore reduce the stated altitude by one-ninth of a degree, leaving, therefore, the true altitude as 11degree53' 20"." The skill and care shown in this record shows that the Arab who observed this eclipse nearly a thousand years ago must have been a man of a different type from an ordinary resident at Bagdad in the year 1899. No description is given of the instrument used, but presumably it was some kind of a quadrant. It does not appear why some of the observations were made at Bagdad and some at Cairo. The Bagdad observations commence with an eclipse of the Sun on November 30, 829, and end with an eclipse of the Moon on November 5, 933. The Cairo observations begin with an eclipse of the Sun on December 12, 977, and end with an eclipse of the Sun on January 24, 1004. These statements apply to the 25 observations which Newcomb considered were trustworthy enough to be employed in his researches, but he rejected three as imperfect.

I have broken away from the strict thread of chronological sequence in order to keep together the notes respecting Arabian observations of eclipses. Let us now revert to the European eclipses.

Under the date of A.D. 733, the Anglo-Saxon Chronicle tells us that, "In this year captured Somerton; and the Sun was eclipsed, and all the Sun's disc was like a black shield; and Acca was driven from his bishopric." Johnston suggests that the reference is to an annular eclipse which he finds occurred on August 14, at about 8h. in the morning. In Schnurrer's Chronik der Seuchen (pt. i., ?113, p. 164), it is stated that, "One year after the Arabs had been driven back across the Pyrenees after the battle of Tours, the Sun was so much darkened on the 19th of August as to excite universal terror." It may be that the English eclipse is here referred to, and a date wrong by five days assigned to it by Schnurrer. Humboldt (Cosmos, vol. iv. p. 384, Bohn's ed.) reports this eclipse in an enumeration he gives of instances of the Sun having been darkened.

On May 5, A.D. 840, there happened an eclipse of the Sun which, amongst other effects, is said to have so greatly frightened Louis Le Debonnaire (Charlemagne's son) that it contributed to his death. The Emperor was taken ill at Worms, and having been removed to Ingelheim, an island in the Rhine, near Mayence, died there on June 20. Hind[75] found that this was a total eclipse, and that the northern limit of totality passed about 100 miles south of Worms. The middle of the eclipse occurred at 1h.?5m.m with the Sun at an altitude of 57? The duration of the eclipse was unusually long, namely about 5?minutes. With the Sun so high and the obscuration lasting so long, this eclipse must have been an unusually imposing one, and well calculated to inspire special alarm.

On Oct. 29, 878, in the reign of King Alfred, there was a total eclipse visible at London. The mention of it in the Anglo-Saxon Chronicle is as follows:--"The Sun was eclipsed at 1 hour of the day." No month is given, and the year is said to have been 879, which is undoubtedly wrong. Hind found that the central line of the eclipse passed about 20 miles N. of London, and that the totality lasted 1m.?1s. Tycho Brahe in his Historia Coelestis quotes from the Annales Fuldenses a statement that the Sun was so much darkened after the 9th hour that the stars appeared in the heavens.

Thorpe in his edition of the Anglo-Saxon Chronicle quotes from Mr. Richard Price a note which assigns the date of March 14, 880, to this eclipse, and cites in confirmation a passage from the Chronicle of Florence of Worcester, anno 879. The 880 eclipse is mentioned by Asser in his De Vit?et Rebus gestis Alfredi in the words following:--"In the same year [879] an eclipse of the Sun took place between three o'clock and the evening, but nearer three o'clock." The confusion of dates is remarkable.

In the Chronicon Scotorum, under the date of 885, we find:--"An eclipse of the Sun; and stars were seen in the heavens." The reference appears to be to the total eclipse of June 16, A.D. 885. The totality lasted more than four minutes, and as the stars are said to have been visible in the North of Ireland, doubtless that part of Ireland came within the eclipse limits.

On Dec. 22, 968, there was an eclipse of the Sun, which was almost total at London at about 8h.?3m.m., or soon after sunrise. The central line passed across the S.-W. of England, and thence through France to the Mediterranean. One Leon, a deacon at Corfu, observed this eclipse, and has left behind what probably is the first perfectly explicit mention of the Corona.[76]

On Aug. 30, 1030, there happened an eclipse visible in Norway, which has already been alluded to on a previous page under the name of the "eclipse of Stiklastad." This was one of those eclipses, the circumstances of which were examined many years ago in detail by Sir G. Airy,[77] because he thought that information of value might be obtained therefrom with respect to the motions of the Moon. Its availability for that purpose has, however, been seriously questioned by Professor Newcomb. Stiklastad is a place where a battle was fought, at which Olav, King of Norway, is said to have been killed. While the battle was in progress the Sun was totally eclipsed, and a red light appeared around it. This is regarded as an early record of the Corona, though not the first.[78] Johnston found that the eclipse was nearly total at about 2h. 21m.m

In 1033 there happened on June 29 an eclipse of the Sun, which evidently had many observers, because it is mentioned by many contemporary writers. For instance, the French historian, Glaber,[79] says that "on the 3rd of the Calends of July there was an eclipse from the sixth to the eighth hour of the day exceedingly terrible. For the Sun became of a sapphire colour; in its upper part having the likeness of a fourth part of the Moon." This sufficiently harmonises with Johnston's calculations that about four-fifths of the Sun on the lower side was covered at 10h. 50m. in the morning.

FOOTNOTES:

[Footnote 61: Hist. Rome, Book lv., chap. 22.]

[Footnote 62: Letter in the Times, July 19, 1872.]

[Footnote 63: Hist. Rome, Book lx., chap. 26.]

[Footnote 64: Life of Apollonius of Tyana, Book viii., c. 23.]

[Footnote 65: Plut. Opera Mor. et Phil., vol. xix. p. 682 Ed. Lipsi? 1778.]

[Footnote 66: Ast. Nach, No. 1838, vol. lxxvii. p. 223: March 31, 1871.]

[Footnote 67: Matheseos, Lib. i., cap. 2, p. 5, Basile? 1533.]

[Footnote 68: Histori? Lib. xx., cap. 3, sec. 1.]

[Footnote 69: Epitome Histori?Ecclesiastic? Lib. xii., cap. 8.]

[Footnote 70: Historia Francorum, Lib. ii., cap. 3 (ad fin.).]

[Footnote 71: Mrs. D. Todd, Total Eclipses of the Sun, p. 101.]

[Footnote 72: The Anglo-Saxon Chronicle, vol. ii. p. 14. Ed. B. Thorpe, 1861.]

[Footnote 73: Prof. S. Ockley, History of the Saracens, vol ii. p. 110. Camb. 1757.]

[Footnote 74: It should be stated that prior to the publication of the work in a book form the greater part of the eclipse observations had been published in the Memoires de l'Institut National des Sciences et Arts: Sciences Mathematiques et Physiques, tome ii.]

[Footnote 75: Letter in the Times, July 19, 1872.]

[Footnote 76: J.Schmidt, Ast. Nach., vol. lxxvii. p. 127, Feb. 1, 1871.]

[Footnote 77: Memoirs, R.A.S., vol. xxvi. p. 131, 1858.]

[Footnote 78: J.燦.燹. Dreyer, Nature, vol. xvi. p. 549, Oct. 25, 1877.]

CHAPTER XII.

One of the most celebrated eclipses of medieval times was that of August 2, 1133, visible as a total eclipse in Scotland. It was considered a presage of misfortune to Henry I. and was thus referred to by William of Malmesbury[80]:--

"The elements manifested their sorrow at this great man's last departure from England. For the Sun on that day at the 6th hour shrouded his glorious face, as the poets say, in hideous darkness agitating the hearts of men by an eclipse; and on the 6th day of the week early in the morning there was so great an earthquake that the ground appeared absolutely to sink down; an horrid noise being first heard beneath the surface."

This eclipse is also alluded to in the Anglo-Saxon Chronicle though the year is wrongly given as 1135 instead of 1133 as it certainly was. The Chronicle says:--"In this year King Henry went over sea at Lammas, and the second day as he lay and slept on the ship the day darkened over all lands; and the Sun became as it were a three-night-old Moon, and the stars about it at mid-day. Men were greatly wonder-stricken and affrighted, and said that a great thing should come hereafter. So it did, for the same year the king died on the following day after St. Andrew's Mass day, Dec. 2, in Normandy." The king did die in 1135, but there was no eclipse of the August new Moon, and without doubt the writer has muddled up the year of the eclipse and of the king's departure from England (to which he never returned) and the year of his death. Calvisius states that this eclipse was observed in Flanders and that the stars appeared.

Respecting the above-mentioned discrepancy Mrs. Todd aptly remarks:--"So Henry must have died in 1133, which he did not; or else there must have been an eclipse in 1135, which there was not. But this is not the only labyrinth into which chronology and old eclipses, imagination, and computation, lead the unwary searcher." Professor Freeman's explanation fairly clears up the difficulty:--"The fact that he never came back to England, together with the circumstances of his voyage, seems to have made a deep impression on men's minds. In popular belief the signs and wonders which marked his last voyage were transferred to the Lammas-tide before his death two years later."[81] The central line of this eclipse traversed Scotland from

Ross to Forfar and the eclipse was of course large in every part of the country. The totality lasted 4m.?0s. in Forfarshire.

Hind has furnished some further information respecting this eclipse. It appears that during the existence of the Kingdom of Jerusalem created by the Crusaders an eclipse occurred which would appear to have been total at Jerusalem or in its immediate neighbourhood. No date is given and a date can only be guessed, and Hind guessed that the eclipse of 1133 was the one referred to. He found that after leaving Scotland and crossing Europe the central line of the 1133 eclipse entered Palestine near Jaffa and passed over Jerusalem where the Sun was hidden for 4?minutes at about 3h.m From Nablous on the N. to Ascalon on the S. the country was in darkness for nearly the same period of time. The alternative eclipses to this one would be those of Sept. 4, 1187, magnitude at Jerusalem 9/10ths of the Sun's diameter; or June 23, 1191, magnitude at the same place about 7/10ths; but these do not seem to harmonise so well with the accounts handed down to us as does the eclipse of 1133.

In 1140, on March 20, there happened a total eclipse of the Sun visible in England which is thus referred to by William of Malmesbury[82]:-- "During this year, in Lent, on the 13th of the Calends of April, at the 9th hour of the 4th day of the week, there was an eclipse, throughout England, as I have heard. With us, indeed, and with all our neighbours, the obscuration of the Sun also was so remarkable, that persons sitting at table, as it then happened almost everywhere, for it was Lent, at first feared that Chaos was come again: afterwards, learning the cause, they went out and beheld the stars around the Sun. It was thought and said by many, not untruly, that the King [Stephen] would not continue a year in the government."

The same eclipse is also thus mentioned in the Anglo-Saxon Chronicle:-- "Afterwards in Lent the Sun and the day darkened about the noontide of the day, when men were eating, and they lighted candles to eat by; and that was the 13th of the Calends of April, March 20. Men were greatly wonder-stricken." The greatest obscuration at London took place at 2h.?6m.m, but it is not quite clear whether the line of totality did actually pass over London.

It was long supposed that this eclipse was total at London, an idea which seems to have arisen from Halley having told the Royal Society anent the total eclipse of May 3, 1715, that he could not find that any total eclipse had been visible at London since March 20, 1140. In consequence of this statement of Halley's, Hind carefully investigated the circumstances of this eclipse, and found that it had not been total at London. The central line entered our island at Aberystwith, and passing near Shrewsbury, Stafford, Derby, Nottingham, and Lincoln, reached the German Ocean, 10 miles S. of Saltfleet. The southern limit of the zone of totality passed through the South Midland counties, and the nearest point of approach to London was a point on the borders of Northamptonshire and Bedfordshire. For a position on the central line near Stafford, Hind found that the totality began at 2h.?6m.m local mean time, the duration being 3m.?6s., and the Sun's altitude being more than 30? The stars seen were probably the planets Mercury and Venus, then within a degree of each other, and 10degreeW. of the Sun, and perhaps the stars forming the well-known "Square of Pegasus." Mars and Saturn were also, at that time, within a degree of each other, but very near the western horizon. It is therefore necessary to look further back than 1140 to find a total solar eclipse visible in London.[83]

A solar eclipse seems to have been alluded to by certain historians as having happened in A.D. 1153. We have the obscure statement that "something singular happened to the Sun the day after the Conversion of St. Paul." A somewhat large eclipse having been visible at Augsburg in Germany, on January 26, this may have been the "something" referred to. It would seem that about 11/12ths of the Sun's diameter was covered.

On May 14, A.D. 1230, there happened a great eclipse of the Sun, thus described by Roger of Wendover[84]:--"On the 14th of May, which was the Tuesday in Rogation Week, an unusual eclipse of the Sun took place very early in the morning, immediately after sunrise; and it became so dark that the labourers, who had commenced their morning's work, were obliged to leave it, and returned again to their beds to sleep; but in about an hour's time, to the astonishment of many, the Sun regained its usual brightness." This eclipse, as regards its total phase, is said by Johnston to have begun in the horizon, a little to the N. of London, in the early morning.

On June 3, A.D. 1239, and October 6, 1241, there occurred total eclipses of the Sun, which have been very carefully discussed by Professor Celoria of Milan, with the view of using them in investigations into the Moon's mean motion.[85] The second of these eclipses is mentioned by Tycho Brahe.[86] He states that "a few stars appeared about noonday, and the Sun was hidden from sight in a clear sky." The eclipse was total in Eastern Europe.

Dr. Lingard,[87] the well-known Roman Catholic historian, speaking of the battle of Cressy, which was fought on August 26, 1346, says:--"Never, perhaps, were preparations for battle made under circumstances so truly awful. On that very day the Sun suffered a partial eclipse: birds in clouds, precursors of a storm, flew screaming over the two armies; and the rain fell in torrents, accompanied with incessant thunder and lightning. About 5 in the afternoon, the weather cleared up, the Sun in full splendour darted his rays in the eyes of the enemy; and the Genoese, setting up their shouts, discharged their quarrels." This was not an eclipse, for none was due to take place; and the phenomenon could only have been meteorological--dense clouds or something of that sort in the sky.

On June 16, 1406, there was a large eclipse of the Sun, 9/10ths of its diameter being covered at London; but on the Continent it seems to have been total. It is stated that the darkness was such that people could hardly recognise one another.

One of the most celebrated eclipses during the Middle Ages was undoubtedly that of June 17, 1433. This was long remembered in Scotland as the "Black Hour," and its circumstances were fully investigated some years ago by Hind. It was a remarkable eclipse in that the Moon was within 13?of perigee and the Sun only 2?from apogee. The central line traversed Scotland in a south-easterly direction from Ross to Forfar, passing near Inverness and Dundee. Maclaurin[88] who lived in the early part of the last century mentions that in his time a manuscript account of this eclipse was preserved in the library of the University of Edinburgh wherein the darkness is said to have come on at about 3 p.m., and to have been very profound. The duration of the totality at Inverness was 4m.?2s.; at Edinburgh 3m.?1s. The central line passed from Britain to the N. of Frankfort-on-the-Maine, through Bavaria, to the Dardanelles, to the S. of Aleppo and thence nearly parallel to the river Euphrates to the N.-E. border of Arabia. In Turkey, according to Calvisius, "near evening the light of the Sun was so overpowered that darkness covered the land."

In 1544, on Jan. 24, there occurred an eclipse of the Sun which was nearly but not quite total. The chief interest arises from the fact that it was one of the first observed by professed astronomers: Gemma Frisius saw it at Louvain.

Kepler says[89] that the day became dark like the twilight of evening and that the birds which from the break of day had been singing became mute. The middle of the eclipse was at about 9 a.m.

In 1560 an eclipse of the Sun took place which was total in Spain and Portugal. Clavius who observed it at Coimbra says[90] that "the Sun remained obscured for no little time: there was darkness greater than that of night, no one could see where he trod and the stars shone very brightly in the sky: the birds moreover, wonderful to say, fell down to the ground in fright at such startling darkness." Kepler is responsible for the statement that Tycho Brahe did not believe this, and wrote to Clavius to that effect 40 years afterwards.

In 1567 there was an annular eclipse visible at Rome on April 9. Clavius says[91] that "the whole Sun was not eclipsed but that there was left a bright circle all round." This in set terms is a description of an annular eclipse, but Johnston who calculated that at Rome the greatest obscuration took place at 20m. past noon points out that the augmentation of the Moon's semi-diameter would almost have produced totality. Tycho tells us that he saw this eclipse on the shores of the Baltic when a young man about 20 years of age.

The total eclipse of February 25, 1598, long left a special mark on the memories of the people of Scotland. The day was spoken of as "Black Saturday." Maclaurin states[92]:--"There is a tradition that some persons in the North lost their way in the time of this eclipse, and perished in the snow"--a statement which Hind discredits. The central line passed from near Stranraer, over Dalkeith, and therefore Edinburgh was within the zone of totality. Totality came on at Edinburgh at 10h.?5m. and lasted 1m.?0s. From the rapid motion of the Moon in declination, the course of the central line was a quickly ascending one in latitude on the Earth's surface, the totality passing off within the Arctic circle.

Kepler in his account of the new star in the constellation Ophiuchus[93] refers to the total eclipse of the Sun of October 12, 1605, as having been observed at Naples, and that the "Red Flames" were visible as a rim of red light round the Sun's disc: at least this seems to be the construction which may fairly be put upon the Latin of the original description.

The partial eclipse of the Sun of May 30, 1612, is recorded to have been seen "through a tube." No doubt this is an allusion to the newly-invented instrument which we now call the telescope. Seemingly this is the first eclipse of the Sun so observed, but it is on record that an eclipse of the Moon had been previously observed through a telescope. This was the lunar eclipse of July 6, 1610, though the observer's name has not been handed down to us.

The eclipse of April 8, 1652, is another of those Scotch eclipses, as we may call them, which left their mark on the people of that country. Maclaurin[94] speaks of it in his time (he died in 1746) as one of the two central eclipses which are "still famous among the populace in this country" [Scotland], and "known amongst them by the appellation of Mirk Monday." The central line passed over the S.E. of Ireland, near Wexford and Wicklow, and reaching Scotland near Burrow Head in Wigtownshire, and passing not far from Edinburgh, Montrose and Aberdeen, quitted Scotland at Peterhead. Greenock and Elgin were near the northern limit of the zone of totality, and the Cheviots and Berwick upon the southern limit. The eclipse was observed at Carrickfergus by Dr. Wyberd.[95] Hind found that its duration there was but 44s. This short duration, he suggested, may partly explain the curious remark of Dr. Wyberd that when the Sun was reduced to "a very slender crescent of light, the Moon all at once threw herself within the margin of the solar disc with such agility that she seemed to revolve like an upper millstone, affording a pleasant spectacle of rotatory motion." Wyberd's further description clearly applies to the Corona. A Scotch account says that "the country people tilling, loosed their ploughs. The birds dropped to the ground."

The eclipse of November 4, 1668, visible as a partial one in England, was of no particular interest in itself but deserves notice as having been observed by Flamsteed,[96] who gives a few diagrams of his observations at Derby. He states that the eclipse came on much earlier than had been predicted. It was well known at this time that the tables of the Sun and Moon then in use were very defective, and it was a recognition of this fact which eventually led to the foundation of the Greenwich Observatory in 1675.

On September 23, 1699, an eclipse of the Sun occurred which was total to the N. of Caithness for the very brief space of 10-15 secs. At Edinburgh, about 11/12ths of the Sun's diameter was obscured. In the Appendix to Pepys's Diary[97] there is a letter from Dr. Wallis mentioning that his daughter's attention was called to it by noticing "the light of the Sun look somewhat dim" at about 9 a.m., whilst she was writing a letter, she knowing nothing of the eclipse.

An eclipse of the Sun occurred on May 12, 1706, which was visible as a partial eclipse in England and was total on the Continent, especially in Switzerland. A certain Captain Stannyan who made observations at Berne, writes thus to Flamsteed[98]:--"That the Sun was totally darkened there for four and a half minutes of time; that a fixed star and a planet appeared very bright; and that his getting out of his eclipse was preceded by a blood-red streak of light from its left limb, which continued not longer than six or seven seconds of time; then part of the Sun's disc appeared all of a sudden as bright as Venus was ever seen in the night; nay, brighter; and in that very instant gave a light and shadow to things as strong as the Moon uses to do."

On this communication Flamsteed remarks:--"The Captain is the first man I ever heard of that took notice of a red streak preceding the emersion of the Sun's body from a total eclipse, and I take notice of it to you [the Royal Society], because it infers that the Moon has an atmosphere; and its short continuance, if only six or seven seconds' time, tells us that its height was not more than five or six hundredths part of her diameter."

On the whole, perhaps, the most celebrated eclipse of the Sun ever recorded in England was that of May 3, 1715. The line of totality passed right across England from Cornwall to Norfolk, and the phenomenon was carefully observed and described by the most experienced astronomer of the time, Dr. Edmund Halley. The line of totality passed over London amongst other places, and as the maximum phase took place soon after 9 a.m. on a fine spring morning, the inhabitants of the Metropolis saw a sight which their successors will not see again till many generations have come and gone. Halley has left behind him an exceedingly interesting account of this event, some allusions to which have already been made.

He seems to have seen what we call the Corona, described by him however as a "luminous ring," "of a pale whiteness, or rather pearl colour, a little tinged with the colours of the Iris, and concentric with the Moon." He speaks also of a dusky but strong red light which seemed to colour the dark edge of the Moon just before the Sun emerged from totality. Jupiter, Mercury, Venus, and the stars Capella and Aldebaran were seen in London, whilst N. of London, more directly under the central line, as many as twenty stars were seen.

The inhabitants of England who lived in the reign of George I. were singularly fortunate in their chances of seeing total eclipses of the Sun, for only nine years after[99] the one just described, namely, on May 22, 1724, another total eclipse occurred. The central line crossed some of the southern countries, and the phenomenon was well seen and reported on by Dr. Stukeley,[100] who stationed himself on Haraden Hill, near Salisbury. The Doctor says of the darkness that he seemed to "feel it, as it were, drop upon us ... like a great dark mantle," and that during the totality the spectacle presented to his view "was beyond all that he had ever seen or could picture to his imagination the most solemn." He could with difficulty discern the faces of his companions which had a ghastly startling appearance. When the totality was ending there appeared a small lucid spot, and from it ran a rim of faint brightness. In about 3?minutes from this appearance the hill-tops changed from black to blue, the horizon gave out the grey streaks previous to morning dawn, and the birds sprang joyously into the air.

This eclipse seems to have had royal observers. It was watched at Kensington apparently by the King or some of the royal family of England, and at Trianon (Paris) by the King of France,[101] under the competent guidance of Maraldi, Cassini and De Louville. It was the last which was visible as a total one in any part of England.

On May 2, 1733, there was an eclipse of the Sun, which was total in Sweden and partial in England. In Sweden the total obscuration lasted more than 3 minutes. Jupiter, the stars in Ursa Major, Capella, and several other stars were visible to the naked eye, as also was a luminous ring round the Sun. Three or four spots of reddish colour were also perceived near the limb of the Moon, but not in immediate contact with it. These so-called red "spots" were doubtless the Red Flames of the present century, and the luminous ring the Corona.

On March 1, 1737, a good annular eclipse was observed at Edinburgh by Maclaurin.[102] In his account he says:--"A little before the annulus was complete a remarkable point or speck of pale light appeared near the middle of the part of the Moon's circumference that was not yet come upon the disc of the Sun.... During the appearance of the annulus the direct light of the Sun was still very considerable, but the places that were shaded from his light appeared gloomy. There was a dusk in the atmosphere, especially towards the N. and E. In those chambers which had not their lights westwards the obscurity was considerable. Venus appeared plainly, and continued visible long after the annulus was dissolved, and I am told that other stars were seen by some." Lord Aberdour mentions a narrow streak of dusky red light on the dark edge of the Moon immediately before the ring was completed, and after it was dissolved. No doubt this is a record of the "Red Flames."

In 1748 Scotland was again favoured with a central eclipse, but it was only annular. The Earl of Morton[103] and James Short, the optician, who observed the phenomenon at Aberdour Castle, 10 miles N.-W. of Edinburgh, just outside the line of annularity, saw a brown coloured light stretching along the circumference of the Moon from each of the cusps. A "star" (probably the planet Venus) was seen to the E. of the Sun.

The annular eclipse of April 1, 1764, visible as such in North Kent, was the subject of the following quaint letter by the Rev. Dr. Stukeley:--

"To the Printer of Whitehall Evening Post,--

"In regard to the approaching solar eclipse of Sunday, April 1, I think it advisable to remark that, it happening in the time of divine service, it is desired you would insert this caution in your public paper. The eclipse begins soon after 9, the middle a little before 11, the end a little after 12. There will be no total darkness in the very middle, observable in this metropolis, but as people's curiositys will not be over with the middle of the eclipse, if the church service be ordered to begin a little before 12, it will properly be morning prayer, and an uniformity preserved in our duty to the Supreme Being, the author of these amazing celestial movements,-- Yours, RECTOR OF ST. GEO., Q.S."[104]

The year 1766 furnishes the somewhat rare case of a total eclipse of the Sun observed on board ship on the high seas. The observers were officers of the French man-of-war the Comte d'Artois. Though the total obscuration lasted only 53 secs., there was seen a luminous ring about the Moon which had four remarkable expansions, situate at a distance of 90?from each other.[105] These expansions are doubtless those rays which we now speak of as "streamers" from the Corona.

Curiously enough the next important total eclipse deserving of notice was also observed at sea. This was the eclipse of June 24, 1778. The observer was the Spanish Admiral, Don Antonio Ulloa, who was passing from the Azores to Cape St. Vincent. The total obscuration lasted 4 minutes. The luminous ring presented a very beautiful appearance: out of it there issued forth rays of light which reached to the distance of a diameter of the Moon. Before it became very conspicuous stars of the 1st and 2nd magnitudes were distinctly visible, but when it attained its greatest brilliancy, only stars of the 1st magnitude could be perceived. "The darkness was such that persons who were asleep and happened to wake, thought that they had slept the whole evening and only waked when the night was pretty far advanced. The fowls, birds, and other animals on board took their usual position for sleeping, as if it had been night."[106]

On Sept. 5, 1793, there happened an eclipse which, annular to the N. of Scotland, was seen and observed in England by Sir W. Herschel[107] as a partial eclipse. He made some important observations on the Moon on this occasion measuring the height of several of the lunar mountains. Considerations respecting the shape of one of the Moon's horns led him to form an opinion adverse to the idea that there the Moon had an atmosphere.

FOOTNOTES:

[Footnote 79: Historiarum Sui Temporis, Lib. iv., cap. 9.]

[Footnote 80: Historia Novella, Lib. i., sec. 8.]

[Footnote 81: Norman Conquest, vol. v. p. 239.]

[Footnote 82: Historia Novella, Lib. ii., sec. 35.]

[Footnote 83: Letter in the Times, July 28, 1871.]

[Footnote 84: Rogerus de Wendover, Flores Historiarum, vol. ii. p. 535, Bohn's ed.]

[Footnote 85: Sugli Eclissi Solari Totali del 3 Giugno 1239, e del 6 Ottobre 1241 in the Memorie del R. Istituto Lombardo di Scienze e Littere, vol. xiii. p. 275.]

[Footnote 86: Historia Coelestis, vol. i. p. 38.]

[Footnote 87: Hist. Engl., vol. iii. chap. xviii. p. 50, 4to. ed.]

[Footnote 88: Phil. Trans., vol. xl. p. 194, 1737.]

[Footnote 89: Astronomi?Pars Optica, c. viii. sec. 3; Opera Omnia, vol. ii. p. 315. Ed. Frisch, 1859.]

[Footnote 90: Quoted by Kepler, as above, at p. 315.]

[Footnote 91: Commentarius in Sacroboscum, cap. iv.; quoted in Kepler's Astronomi?Pars Optica, c. viii. sec. 3; Opera Omnia, vol ii. p. 316. Ed. Frisch, 1859.]

[Footnote 92: Phil. Trans., vol. xl. p. 193; 1737.]

[Footnote 93: De Stell?Nov?in Pede Serpentarii, p. 115; Prag? 1606.]

[Footnote 94: Phil. Trans., vol. xl. p. 193; 1737.]

[Footnote 95: V. Wing, Astronomia Britannica, p. 355.]

[Footnote 96: Historia Coelestis, vol. i. pp. 7 and 21.]

[Footnote 97: Diary of Samuel Pepys, vol. vi. p. 208; Ed. M. Bright, 1879.]

[Footnote 98: Phil. Trans., vol. xxv. p. 2240; 1706.]

[Footnote 99: Being half a Saros period (see p. 20, ante).]

[Footnote 100: Itinerarium Curiosum, 2nd ed., vol. i. p. 180.]

[Footnote 101: Mem. de Mathematique et de Physique de l'Acad. des Sciences, 1724, p. 259.]

[Footnote 102: Phil. Trans., vol. xl. pp. 181, 184. 1737.]

[Footnote 103: Phil. Trans., vol. xlv. p. 586. 1750. This is the man who under the designation of "Lord Aberdour" observed the eclipse of 1737 (ante).]

[Footnote 104: Rev. W. Stukeley, Rector of St. George's, Queen's Square, London, Diary, vol. xx. p. 44, ed. "Surtees Soc.," vol. lxxvi. p. 384.]

[Footnote 105: Le Gentil, Voyage dans les Mers de l'Inde, vol. ii. p. 16. Paris, 1769.]

[Footnote 106: Phil. Trans., vol. lxix. p. 105. 1779.]

[Footnote 107: Phil. Trans., vol. lxxxiv. p. 39. 1794.]

CHAPTER XIII.

ECLIPSES OF THE SUN DURING THE NINETEENTH CENTURY.

Observations of total solar eclipses during the 19th century have been, for the most part, carried on under circumstances so essentially different from everything that has gone before, that not only does a new chapter seem desirable but also new form of treatment. Up to the beginning of the 18th century the observations (even the best of them) may be said to have been made and recorded with but few exceptions by unskilled observers with no clear ideas as to what they should look for and what they might expect to see. Things improved a little during the 18th century and the observations by Halley, Maclaurin, Bradley, Don Antonio Ulloa, Sir W. Herschel, and others in particular rose to a much higher standard than any which had preceded them. However, it has only been during the 19th century, and especially during the latter half of it, that total eclipses of the Sun have been observed under circumstances calculated to extract from them large and solid extensions of scientific knowledge. Inasmuch as it has been deemed convenient to sort out and classify our knowledge under particular heads in previous chapters, I shall in this chapter speak only of the leading facts of each eclipse in such an outline form as will avoid as far as possible unnecessary repetition.

In 1806 a total eclipse of the Sun occurred, visible in N. America. Observations made in the United States have been handed down to us. Don Joachin Ferrer, a Spanish astronomer, observed the eclipse at Kinderhook in the State of New York. The totality lasted more than 4 m.--a somewhat unusual length of time. One or two planets and a few 1st magnitude stars were seen. During the totality there was a slight fall of dew.

On Nov. 19, 1816, there occurred the first total eclipse of the Sun in the 19th century, the central line of which passed over Europe. There is only one known observation of the total phase, and this was by Hagen at Culm in Bohemia, but he appears to have seen only the beginning of the totality and not the whole of it.

A partial eclipse of the Sun visible as such in England but which was annular in the Shetland Isles took place on Sept. 7, 1820. The only reason why this is worth mention is for its political associations. The trial of Queen Caroline was going on in the House of Lords, and the House suspended its sitting for a short time for the sake of the eclipse.

On May 15, 1836, there occurred an annular eclipse of the Sun, which though it was nowhere total, may be looked upon as the first of the modern eclipses the observations of which have taken such a great development during recent years. The annularity of this eclipse was observed in the N. of England and in the S. of Scotland; and it was at Jedburgh in Roxburghshire that Mr. Francis Baily[108] observed that feature of eclipses of the Sun now universally known as "Baily's Beads." Some indications of the Red Flames were also obtained at places where the eclipse was annular.

Probably it was the recognition of Baily's Beads as a regular concomitant of eclipses of the Sun, which helped to pave the way for the extensive preparations made in France, Italy, Austria, and Russia for observing the total eclipse of July 8, 1842. Many of the most eminent astronomers of Europe repaired to different stations on the central line in order to see the phenomenon. Amongst these may be named Arago, Valz, Airy, Carlini, Santini, and O. Struve. The eclipse was witnessed under favourable circumstances at all the various stations on the central line across Europe, from Perpignan in France in the West to Lipesk in Russia in the East.

Arago wrote[109] such an exceedingly graphic account of this eclipse from what may be termed the standpoint of the general public, that I will quote it at some length, because, with an alteration of date, it might be re-written and applied to every total eclipse visible in much populated tracts of country.

"At Perpignan persons who were seriously unwell alone remained within doors. As soon as day began to break the population covered the terraces and battlements of the town, as well as all the little eminences in the neighbourhood, in hopes of obtaining a view of the Sun as he ascended above the horizon. At the citadel we had under our eyes, besides numerous groups of citizens established on the slopes, a body of soldiers about to be reviewed.

"The hour of the commencement of the eclipse drew nigh. More than twenty thousand persons, with smoked glasses in their hands, were examining the radiant globe projected upon an azure sky. Although armed with our powerful telescopes, we had hardly begun to discern the small notch on the western limb of the Sun, when an immense exclamation, formed by the blending together of twenty thousand different voices, announced to us that we had anticipated by only a few seconds the observation made with the unaided eye by twenty thousand astronomers equipped for the occasion, whose first essay this was. A lively curiosity, a spirit of emulation, the desire of not being outdone, had the privilege of giving to the natural vision an unusual power of penetration. During the interval that elapsed between this moment and the almost total disappearance of the Sun we remarked nothing worthy of relation in the countenances of so many spectators. But when the Sun, reduced to a very narrow filament, began to throw upon the horizon only a very feeble light, a sort of uneasiness seized upon all; every person felt a desire to communicate his impressions to those around him. Hence arose a deep murmur, resembling that sent forth by the distant ocean after a tempest. The hum of voices increased in intensity as the solar crescent grew more slender; at length the crescent disappeared, darkness suddenly succeeded light, and an absolute silence marked this phase of the eclipse with as great precision as did the pendulum of our astronomical clock. The phenomenon in its magnificence had triumphed over the petulance of youth, over the levity which certain persons assume as a sign of superiority, over the noisy indifference of which soldiers usually make profession. A profound stillness also reigned in the air; the birds had ceased to sing. After an interval of solemn expectation, which lasted about two minutes, transports of joy, shouts of enthusiastic applause, saluted with the same accord, the same spontaneous feeling, the first reappearance of the rays of the Sun. To a condition of melancholy produced by sentiments of an indefinable nature there succeeded a lively and intelligible feeling of satisfaction which no one sought to escape from or moderate the impulses of. To the majority of the public the phenomenon had arrived at its term. The other phases of the eclipse had few attentive spectators beyond the persons devoted especially to astronomical pursuits."

The total eclipse of July 28, 1851, may be said to have been the first which was the subject of an "Eclipse Expedition," a phrase which of late years has become exceedingly familiar. The total phase was visible in Norway and Sweden, and great numbers of astronomers from all parts of Europe flocked to those countries. Amongst those who went from England were Sir G. Airy, the Astronomer Royal (then Mr. Airy), Mr. J. Hind and Mr. Lassell. The Red Flames were very much in evidence, and the fact that they belonged to the Sun and not to the Moon was clearly established. Hind mentions that "the aspect of Nature during the total eclipse was grand beyond description." This feature is dwelt upon with more than usual emphasis in many of the published accounts. I have never seen it suggested that the mountainous character of the country may have had something to do with it, but that idea would seem not improbable.

In the year 1858, two central eclipses of the Sun occurred, both presenting some features of interest. That of March 15 was annular, the central line passing across England from Lyme Regis in Dorsetshire to the Wash, traversing portions of Somersetshire, Wiltshire, Berkshire, Oxfordshire, Northamptonshire, Lincolnshire, and Norfolk. The weather generally was unfavourable and the annular phase was only observed at a few places, but important meteorological observations were made and yielded results, as regards the diminution of temperature, which were very definite. All over the country rooks and pigeons were seen returning home during the greatest obscuration; starlings in many places took flight; at Oxford a thrush commenced its evening song; at Ventnor a fish in an aquarium, ordinarily visible in the evening only, was in full activity about the time of greatest gloom; and generally, it was noted that the birds stopped singing and flew low from bush to bush. The darkness, though nowhere intense, was everywhere very appreciable and decided. The second central eclipse of 1858 took place on September 7 and was observed in Peru by Lieutenant Gilliss of the U.S. Navy. The totality only lasted one minute, and the general features of a total eclipse do not appear to have been very conspicuously visible. Gilliss remarks[110]:--"Two citizens of Olmos stood within a few feet of me, watching in silence, and with anxious countenances, the rapid and fearful decrease of light. They were wholly ignorant that any sudden effect would follow the total obscuration of the Sun. At that instant one exclaimed in terror "La Gloria," and both, I believe, fell to their knees, filled with awe. They appreciated the resemblance of the Corona to the halos with which the old masters have encircled their ideals of the heads of our Saviour and the Madonna, and devoutly regarded this as a manifestation of the Divine Presence."

The year 1860 saw the departure from England of the first great Ship Expedition to see an eclipse. One was due to happen on July 18, and a large party went out from England to Spain in H.M.S. Himalaya. Mr. De La Rue took a very well-equipped photographic detachment, and his photographs were eminently successful. This eclipse settled for ever the doubt as to whether the Red Flames belonged to the Sun or the Moon, and in favour of the former view.

The years 1868, 1869, and 1870 were each marked by total eclipses, which were observed to a greater or less extent. In the first-named year the eclipse occurred on August 18, the central line passing across India. The weather was not everywhere favourable, but several expeditions were dispatched to the East Indies. The spectroscope was largely brought into play with the immediate result of showing that the Corona was to be deemed a sort of atmosphere of the Sun, not self-luminous, but shining by reflected light. The eclipse of 1869 was observed by several well-equipped parties in the United States, and a very complete series of excellent photographs was obtained.

To view the eclipse of December 22, 1870, several expeditions were dispatched, the central line passing over some very accessible places in Spain, Sicily, and North Africa. The English observers went chiefly in H.M.S. Urgent, though some of them travelled overland to Sicily. The expenses, both of the sea and land parties, were to a large extent defrayed by Her Majesty's Government. It deserves to be noted that so great was the anxiety of the French astronomer Janssen to see this eclipse, that he determined to try and escape in a balloon from Paris (then besieged by the Germans) and succeeded, carrying his instruments with him. The weather seriously interfered with the work of all the observers who went out to see this eclipse, which was the more to be regretted because the preparations had been on a very extensive and costly scale. The chief result was that it was ascertained that the Red Flames (hence forward generally called "Prominences") are composed of hydrogen gas in an incandescent state.

The year 1871 saw, on December 12, another Indian eclipse, noteworthy for the numerous and excellent photographs which were obtained of the Corona, of the rifts in it, and of the general details, which were well recorded on the plates.

There was an eclipse visible in South Africa on April 16, 1874. Some useful naked eye views were obtained and recorded, but as no photographic work was done, this eclipse cannot be said to come into line with those which preceded or followed it.

In the following year, that is to say on April 6, 1875, there was a total eclipse of the Sun, visible in the far East, especially Siam; but the distance from England, coupled with the very generally unfavourable weather, prevented this from being anything more than a second-class total eclipse, so to speak, although extensive preparations had been made, and the sum of ?000 had been granted by the British Government towards the expenses. A certain number of photographs were obtained, but none of any very great value.

Perhaps of the next eclipse which we have to consider, it may be said that the circumstances were more varied than those of any other during the second half of the 19th century. The eclipse in question occurred on July 29, 1878.

Several favourable circumstances concurred to make it a notable event. In the first place, the central line passed entirely across the United States; in other words, across a long stretch of inhabited and civilised territory, accessible from both sides to a nation well provided with the requisite scientific skill and material resources of every kind. But there was another special and rare facility available: the central line crossed the chain of the Rocky Mountains, an elevated locality, which an American writer speaks of as overhung by "skies of such limpid clearness, that on several evenings Jupiter's satellites were seen with the naked eye." On the summit of a certain peak, known as Pike's Peak, a party of skilled observers, headed by Professor Langley, observed the wonderful developments of the Corona, mentioned on a previous page. The fact that such a display came under the eyes of man was no doubt mainly due to the superbly clear atmosphere through which the observations were made. That this is not a mere supposition may be inferred from the fact that at the lower elevation of only 8000 feet, instead of 14,000 feet, the Coronal streamers were seen by Professor Newcomb's party, far less extended than Langley saw them. Perhaps the best proof of the importance of a diaphanous sky is to be found in the fact that on the summit of Pike's Peak, the Corona remained visible for fully 4 minutes after the total phase had come to an end. A comparison of the descriptions shows that even at the elevation of 10,200 ft. the observers placed there, whilst they were better off than those at 8000 ft., assuredly did not see so much or so well as those at 14,000 ft.

There occurred a total eclipse on July 11, 1880, visible in California, but as the totality lasted only 32 secs. and the Sun's elevation was only 11? not much was got out of this eclipse notwithstanding that it was observed in a cloudless sky at a station 6000 ft. above the sea.

The eclipse of May 17, 1882, yielded several interesting and important features although the totality was short--only about 1?minutes. Here again favourable local circumstances helped astronomers in more ways than one. It was in Egypt that the eclipse was visible, and Egypt is a country which it is exceedingly easy for travellers to reach, and it is also noted for its clear skies. These were doubtless two of the reasons which combined to inspire the elaborate preparations which were made for photographic and spectroscopic observations. The former resulted in a very unprecedented success. One of Dr. Schuster's photographs of the totality showed not only the expected Corona, but an unexpected comet.

Though on more than one previous occasion in history the darkness which is a special accompaniment of a total eclipse had caused a comet to be seen, yet the 1882 eclipse was the first at which a comet had thrust itself upon the notice of astronomers by means of a photographic plate. It will be remembered that the political circumstances of Egypt in 1882 were of a somewhat strained character and probably this contributed to the development of an unusual amount of astronomical competition in connection with this eclipse. Not only did the Egyptian Government grant special facilities, but strong parties went out representing England, France, and Italy, although not perhaps in set terms at the direct instigation of their respective Governments.

The next eclipse, that of May 6, 1883, had some dramatic features about it. To begin with its duration was unusually long--nearly 5?minutes, and Mrs. Todd in her genial American style remarks:--"After the frequent manner of its kind, the path lay where it would be least useful--across the wind-swept wastes of the Pacific. But fortunately one of a small group of coral islands lay quite in its line, and, nothing daunted, the brave scientific men set their faces toward this friendly cluster, in cheerful faith that they could locate there. Directed to take up their abode somewhere on a diminutive island about which nothing could be ascertained beforehand, save the bare fact of its existence at a known spot in mid-ocean, the American observers were absent from the United States more than three months, most of which time was spent in travelling, 15,000 miles in all, with ten full weeks at sea. Their tiny foothold in the Pacific was Caroline Island, a coral atoll on the outskirts of the Marquesas group."

In spite of the unattractive, not to say forbidding, character of the place to which they would have to go, parties of astronomers went out from England, France, Austria, and Italy, and although rain fell on the morning of the day the sky became quite clear by the time of totality and the observations were completely successful. One of the pictures of the Corona obtained by Trouvelot, an observer of French descent, but belonging to the American party, has been often reproduced in books and exhibited the Corona in a striking form. How few were the attractions of Caroline Island as an eclipse station may be judged from the fact that the inhabitants consisted of only four native men, one woman, and two children who lived in three houses and two sheds.

On September 8, 1885, there occurred a total eclipse, which was seen as such in New Zealand, but the observations were few, and with one exception, unimportant and uninteresting. A certain Mr. Graydon, however, made a sketch which showed at one point a complete break in the Corona so that from the very edge of the Moon outwards into space, there was a long and narrow black space showing nothing but a vacuity. If this was really the condition of things, such a break in the Corona is apparently quite unprecedented.

In 1886, on August 29, there occurred a total eclipse, visible in the West Indies, which yielded various important results. It was unfortunate that for the greater part of its length, the zone of totality covered ocean and not land, the only land being the Island of Grenada and some adjacent parts of South America. The resulting restriction as regards choice of observing stations was the more to be regretted because the duration of the totality was so unusually long, and therefore favourable, being more than 6?minutes in the middle of the Atlantic Ocean. Parties of English, American, and Italian astronomers assembled, however, at Grenada, and though the weather was not the best possible, some interesting photographs were obtained which exhibited an unusual development of hydrogen protuberances. The central line in this eclipse not only stretched right across the Atlantic, but entered Africa on the West Coast where a missionary saw the eclipse as a mere spectator, and afterwards expressed his regret that no astronomers were within reach with instruments to record the remarkable Corona which was displayed to his gaze.

Though the unusual opportunities which, so far as the Sun and the Moon were concerned, were afforded by the eclipse of 1886 were lost, astronomers looked out hopefully for August 19, 1887, when another eclipse was due to happen which, weather permitting, would be observable over a very long stretch of land, from Berlin through Russia and Siberia to Japan. Unusually extensive preparations were made in Russia at one end and in Japan at the other, but clouds prevailed very generally, and the pictures of the Corona which were obtained fell far short in number and quality from what had been hoped for, having regard to the number and importance of the stations chosen, and of the astronomers who made their preparations thereat. An enthusiastic Russian, in the hopes of emancipating himself from the risks of terrestrial weather at the Earth's surface, went up in a balloon to an elevation of more than two miles. His enthusiasm was so far rewarded that he had a very clear view of a magnificent Corona; but as, owing to some mischance, the balloon rose, conveying only the astronomer and leaving behind his assistant who was to have managed the balloon, all his time was engrossed by the management of the balloon, and he could do very little in the way of purely astronomical work.

The year 1889 afforded two total eclipses of the Sun for which the usual preparations were made. The first occurred on New Year's Day, and the path of the shadow crossed the North American Continent from California to Manitoba. The weather was nearly everywhere very favourable, and an enormous number of observers and instruments were assembled along the central line. The consequence was that a very large number of photographs were obtained. It may be said generally of this eclipse, that as it coincided with a Sun-spot minimum, it left us in a position to learn very distinctly what are the characteristic features of a solar Corona at a period which is one of rest and repose on the Sun, at least, so far as regards visible Sun-spots.

The second eclipse of 1889 occurred on December 22, and should have been visible off the northern coast of South America and on the West Coast of Africa. Attempts were made to utilise the South American chances by English and American parties, whilst a small expedition comprising astronomers of both nations went to Cape Ledo in West Africa. The African efforts failed entirely owing to clouds, but the South American parties at Cayenne were successful. One very deplorable result, however, arising out of the expedition to Cayenne was the illness and subsequent death of the Rev. S. Perry, S.J., who was struck down by malaria and died at sea on the return journey. None who knew Mr. Perry personally could fail to realise what a loss he was both to astronomy generally and to his own circle of friends particularly.

On April 16, 1893, there happened a total eclipse of the Sun, which was successfully watched by a large number of skilled observers throughout its entire length. Indeed it is believed that only one party was unsuccessful. The line of totality started on the coast of Chili, passed over the highlands of that country, across the borders of Argentina and Paraguay, and over the vast plains and forests of Central Brazil, emerging at about noon of local time at a short distance to the N.-W. of Ceara on the North Atlantic seaboard. Crossing the Atlantic nearly at its narrowest part, it struck the coast of Africa N. of the river Gambia, and finally disappeared somewhere in the Sahara. The South American observations were the most extensive and successful, the latter fact being due to the circumstance that the sky at many of the principal stations was pre-eminently favourable, owing to the clearness and dryness of the atmosphere.

On Sept. 29, 1894, there was a total eclipse of the Sun, but as its duration was brief and the zone of totality lay chiefly over the Indian Ocean, practically nothing came of it.

Things seemed, however, much more promising for the total eclipse of Aug. 9, 1896, and a very large number of observers went out to the North of Norway hoping to catch the shadow at its European end, whilst a yacht party went to Nova Zembla in the Arctic Ocean, and a few observers travelled as far as Japan. So far as the very large number of would-be observers who went from England to Norway were concerned, the eclipse was a profound disappointment, for owing to bad weather practically nothing was seen in Norway except on the West coast near Bod? where the weather was beautifully fine, but where no adequate preparations had been made, because nobody believed that the coast would be free from fog. Exceptionally fine weather prevailed at Nova Zembla, and the small but select party who were kindly taken there by the late Sir G. Powell, M.P., in his yacht, were very fortunate, and an excellent series of photographs was secured. One important result obtained at Nova Zembla was a full confirmation by Mr. Shackleton of Prof. Young's discovery in 1870 of the "Reversing Layer," a discovery which was long and vehemently disputed by Sir Norman Lockyer. Fairly successful observations were made of this eclipse in Siberia and Japan.

The last total eclipse of the Sun which has to be noticed as an accomplished fact was the "Indian Eclipse" of Jan. 22, 1898, which was very successfully seen by large numbers of people who went to India from all parts of the world. As usual in all total eclipses of the Sun nowadays, the photographers were very much to the front, and the photographs of the inner Corona, taken by the Astronomer Royal, are thought to have been probably the best that have yet been done. Amongst the miscellaneous observations made, it may be mentioned that more stars were seen during the second partial phases than during totality (a circumstance which had been noticed by Don A. Ulloa as far back as 1778). It is stated also that a mysterious object was seen between Mars and Venus by two officers of H.M.S. Melpomene, which was not put down on the published chart as a star to be looked for. The identity of this object has not been ascertained.

FOOTNOTES:

[Footnote 108: Memoirs, R.A.S., vol. x. p. 5.]

[Footnote 109: L'Annuaire, 1846, p. 303.]

[Footnote 110: Month. Not., R.A.S., vol. xx. p. 301; May 1860.]

CHAPTER XIV.

THE ELECTRIC TELEGRAPH AS APPLIED TO ECLIPSES OF THE SUN.

Amongst the auxiliary agencies which have been brought into use in recent years, to enable astronomers the better to carry out systematic observations of eclipses of the Sun, the electric telegraph occupies a place which may hereafter become prominent. As it is not likely that this little book will fall into the hands of any persons who would be able to make much use of telegraphy in connection with eclipse observations, it will not be necessary to give much space to the matter, but a few outlines will certainly be interesting. When the idea of utilising the telegraph wire first came into men's minds, it was with the object of enabling observers who saw the commencement of an eclipse at one end of the line of totality, to give cautionary notices to observers farther on, or towards the far end, of special points which had been seen at the beginning of the totality, and as to which confirmatory observations, at a later hour, were evidently very desirable. It is obvious that a scheme of this kind depends for its success upon each end (or something like it) of the line of totality being in telegraphic communication with the other end, and this involves a combination of favourable circumstances not likely to exist at every occurrence of a total eclipse, and in general only likely to prevail in the case of eclipses visible over inhabited territory, such as the two Americas, Europe, and parts of Asia. This use of the telegraph was, I think, first proposed as far back as 1878, by an American astronomer, in connection with the total eclipse of that year. His proposal fell upon sympathetic ears, with the result that arrangements were concluded with the Western Union Telegraph Company of North America for the expeditious forwarding of messages from northern stations on the eclipse line to southern stations. Some attention was being given at that time to the question of Intra-Mercurial planets, and it was thought that if by good fortune any such objects were unexpectedly found at the northern station, and observers at a southern station could be advised of the fact, there might be a better chance of procuring an accurate and precise record of the discovery. As it happened, nothing came of it on that occasion, but the idea of utilising the telegraph having once taken possession of men's minds, it was soon seen what important possibilities were opened up.

The want of telegraph organisation curiously made itself felt in the Egyptian eclipse of 1882. It is stated in another chapter of this work that during the total phase a comet was unexpectedly discovered. Now comets sometimes move very rapidly (especially when they are near the Sun), and had it been possible to have warned some observer to the E. of Egypt to look out for this comet, and had he seen it even only a couple of hours after it had been found in Egypt, some data respecting its position might have been obtained which would have permitted a rough estimate being formed of its movement through the heavens. Such an estimate might have enabled astronomers to have hunted up the comet at sunset or sunrise on the days immediately following the eclipse. As it happened, however, the comet was not seen again in 1882, and, so far as we know, may never be seen again.

It was not till 1889 that a complete organisation of a telegraph service in connection with an eclipse was accomplished. The eclipse of January 1 of that year began in the Pacific and the line of totality touched land in California, passing across North America to Manitoba. The first Californian station was at Willows, and was occupied by a party from Harvard College Observatory, who were supplied with an unusually complete equipment of photographic apparatus, together with a large camera for charting all the stars in the neighbourhood of the Sun, so as to detect an Intra-Mercurial planet if one existed. The telegraph scheme which had to be worked out was somewhat complicated, and one of the chief actors in the scene has furnished a fairly full account of what was done. First of all, a complete list of the instruments and of the work proposed to be done by them had to be prepared. The weather probabilities being everywhere very unsatisfactory, there was a possibility of all degrees of success or failure, and one thing which had to be prearranged for each station was a cypher code which should be available for all the likely combinations of instruments, weather and results. It was found that about one hundred words would suffice for the necessary code, including words which would indicate in a sufficiently precise manner the position of any new planet which a photograph might disclose.

The following, being a part of the code prescribed for use at Willows, will serve to indicate the nature of the whole scheme:--

Africa, Perfectly clear throughout the whole eclipse. Alaska, Perfectly clear during totality. Belgium, Clear sky for the partial phases, but cloudy for totality. Bolivia, Entirely cloudy throughout the whole eclipse. Brazil, Observed all the contacts. Bremen, Observed three of the contacts. Ceylon, Made observations on the shadow-bands. Chili, Observed lines of the reversing layer visually. China, The Corona showed great detail. Cork, Obtained 40-50 negatives during totality. Corsica, Obtained 50-60 negatives during totality. Crimea, Obtained 60-70 negatives during totality. Cuba, Observed a comet.

Upwards of twenty codes were prepared for the like number of stations, and the observers were to report their results at the earliest possible moment. On a rehearsal of the programme the thought occurred that the sending and reception of so many cypher messages in the ordinary course of business might lead to delays which would be productive of serious inconvenience, and that the success of the whole scheme could be only well assured if a special wire, in direct circuit from New York to the eclipse stations in turn, could be dedicated to the work. Thanks to the liberality of the Western Union Telegraph Company this privilege was secured, and a branch wire was led across from the Company's New York office to the office of the New York Herald, which journal had undertaken to be responsible for the non-astronomical part of the business.

Mrs. Todd gives the following account of the final arrangements, and of how they began to work when the moment for action arrived:--"From San Francisco every California observer was within easy telegraphic reach, and the wire thus extended by direct circuit to each eclipse station in turn. From the editorial rooms of the Herald Professor Todd was in immediate communication with any observers whom he chose to call. As previously intimated, arrangements had been made with the Harvard astronomers at Willows to receive their message first and with the utmost despatch, in order to test the feasibility of outstripping the Moon. Shortly before 5 o'clock in the afternoon despatches began to come in. Of course a slight delay was unavoidable, as the observers at the various stations were some rods distant from the local telegraph offices, and it would take a few minutes after the eclipse was over to prepare the suitable message from the cypher code. On the astronomer's table in the Herald office were a large map and a chronometer. The latter indicated exact Greenwich time, and the former showed the correct position of the Moon's shadow at the beginning of every minute by the chronometer. In this way it was possible to follow readily the precise phase of the eclipse at every station. About the rooms and accessible for immediate use were arranged the cypher codes pertaining to the several stations and other papers necessary in preparing the reports for the press. Everything being, as was supposed, in working order, New York about a quarter of an hour before totality commenced inquired of Willows the state of the weather. The answer was that the sky was getting dark, and that there were no clouds anywhere near the Sun. At that time the Moon's shadow was travelling across the open waters of the Pacific. It rapidly rushed along; totality came and went at Willows; a two minutes' glimpse of the Corona was had, and the Corona swept rapidly eastwards. After a brief interval Professor Pickering sent off from Willows a telegram which began--"Alaska, China, Corsica," and then the connection failed. The break was located somewhere between California and Utah, and more than half an hour elapsed ere the circuit was re-established, and the rest of the message received. The remainder of the thrilling incidents of that eventful day cannot possibly be better told than in Mrs. Todd's crisp and striking language[111]:--

"During this interval the lunar shadow, advancing over Montana and Dakota, had left the Earth entirely, sweeping off again into space. Still, however, the prospect that the telegraph might win the race was hopeful. Had New York been located in the eclipse path as well as Willows, with both stations symmetrically placed, the total eclipse would have become visible at New York about an hour and a quarter after the shadow had left California. Thus there was time to spare. Having recovered the wire, Professor Pickering's message was completed at 10h.?6m. [G.M.T.], the cypher translated, and the stenographer's notes were written out and despatched to the composing-room six minutes later. The "copy" was quickly put in type, and the hurried proof handed to Professor Todd at 10h.?0m., exactly an hour of absolute time after the observations were concluded. Had the Moon's shadow been advancing from California toward New York, there was still a margin of several minutes before the eclipse could become total at the latter place. In point of fact, while the proof sheet of the first message was being read, the lunar shadow would have been loitering among the Alleghanies. Man's messenger had thus outrun the Moon. The telegraphic reports of the other astronomers were gradually gathered and put in type, and the forms of the Herald were ready for the stereotyper at the proper time, some two hours after midnight. At 3 o'clock a.m. the European mails closed, and the pouches put on board the steamship Aller carried the usual copies for the foreign circulation. Within twenty-four hours after the observations of the eclipse were made near the Pacific coast, the results had been telegraphed to the Atlantic seaboard, collected and printed, and the papers were well out on their journey to European readers."

The foregoing narrative will make amply clear the future possibilities of telegraphy as a coadjutor of Astronomy in the observation of total eclipses of the Sun. And if the will and the funds are forthcoming, the eclipse of May 28, 1900, will afford an excellent opportunity of again putting to the test the excellent ideas of which our American friends worked out so successfully ten years ago. The zone of totality in that eclipse passing as it will through so many of the densely populated Southern States of North America, and then through Portugal, Spain, and Algiers, great facilities will present themselves for telegraphic combinations, if political and financial difficulties do not interfere.

FOOTNOTES:

[Footnote 111: There is a want of uniformity in Mrs. Todd's references to times which I have not thought it necessary to put straight. "Greenwich Mean Time," "Eastern U.S. Standard Time," and "Pacific Time," are all severally quoted in happy-go-lucky confusion.]

CHAPTER XV.

ECLIPSES OF THE MOON--GENERAL PRINCIPLES.

In dealing with eclipses generally, but with more especial reference to eclipses of the Sun, in a previous chapter, it was unavoidable to mix up in some degree eclipses of the Moon with those of the Sun. There are, however, distinctions between the two phenomena which make it convenient to separate them as much as possible. Eclipses of the Moon are, like those of the Sun, divisible into "partial" and "total" eclipses, but those words have a different application in regard to eclipses of the Moon from what they have when eclipses of the Sun are in question. A little thought will soon make it clear why this should be the case. A partial eclipse of the Sun results from the visible body of the Sun being in part concealed from us by the solid body of the Moon, and so in a total eclipse there is total concealment of the one object by the other.

But when we come to deal with partial and total eclipses of the Moon, the situation, is materially different. The Moon becomes invisible by passing into the dark shadow cast by the Earth into space.

[Illustration: FIG. 13.--THEORY OF AN ECLIPSE OF THE MOON.]

Fig. 13 will make this clear without the necessity of much verbal explanation. S represents the Sun, E the Earth, and mn the orbit of the Moon. It is obvious that whilst the Moon is moving from m to n it becomes immersed in the Earth's shadow. But before actually reaching the shadow the Moon passes through a point in its orbit at which it begins to lose the full light of the Sun. This is the entrance into the "penumbra" (or "Partial shade"). Similarly, after the eclipse, when the Moon has emerged from the full shadow it does not all at once come into full sunshine, but again passes through the stage of penumbral illumination,[112] and under such circumstances (to speak in the style of Old "Oireland") the invisible Moon is very often not invisible, and the part partially eclipsed is often not eclipsed, and when the Moon is totally eclipsed it is frequently still visible. Of course the general idea involved in all cases of a body passing into the shadow of another body is that the body which so passes disappears, because all direct light is cut off from it. In the case, however, of a lunar eclipse this state of things is not always literally accomplished, and very often some residual light reaches the Moon (of course from the Sun) with the result that traces of the Moon may often be discerned. The laws which govern this matter are very ill-understood. The fact remains that if we examine a series of reports of observed eclipses of the Moon extending over many centuries (and records exist which enable us to do this) we shall find that in some instances when the Moon was "totally" eclipsed in the technical sense of that word, it was still perfectly visible, whilst during other eclipses it absolutely and entirely disappeared from view. Such eclipses are sometimes spoken of as "black" eclipses of the Moon, but the phrase is not a happy one. Many instances of both kinds will be found mentioned in the chapter on historical lunar eclipses.[113]

The different conditions of eclipses of the Moon are illustrated by Fig. 14 which must be studied with the aid of the remarks made in a former chapter concerning the apparent movements of the Sun and Moon and their nodal passages. Suffice it to state here that in Fig. 14 AB represents the ecliptic, and CD the Moon's path. The three black circles are imaginary sections of the Earth's shadow as cast when the Earth is in three successive positions in the ecliptic. If when the Earth's shadow is near A the Moon should be at E, and in Conjunction with the Earth the Moon will escape eclipse; if the Conjunction takes place with both the Earth's shadow and the Moon a little further forward, say at F, the Moon will be partially obscured; but if the Moon is at or very near its node, as at G, it will be wholly involved in the Earth's shadow and a total eclipse will be the result. In the case contemplated at G in the diagram, the Moon is concentrically placed with respect to the shadow, but the eclipse will equally be total even though the two bodies are not concentrically disposed, so long as the Moon is wholly within the cone of the Earth's shadow.[114]

Just as in the case of the Sun so with the Moon there are certain limits on the ecliptic within which eclipses of the Moon may take place, other (narrower) limits within which they must take place, and again other limits beyond which they cannot take place. Reverting to what has been said on a previous page[115] with respect to these matters when an eclipse of the Sun is in question it is only necessary to substitute for the word "Conjunction," the word "Opposition"; and for 18degrees and 15 及 of longitude the figures 12degrees and 9 及. The limits in latitude will be 1degree3' and 0degree52' instead of 1degree34' and 1degree23'. These substitutions made, the general ideas and facts stated with regard to the conditions of an eclipse of the Sun will apply also to the one of the Moon.

It is to be noted that whereas eclipses of the Sun always begin on the W. side of the Sun, eclipses of the Moon begin on the E. side of the Moon. This difference arises from the fact that the Sun's movement in the ecliptic is only apparent (it being the Earth which really moves), whilst the Moon's movement is real.

Eclipses of the Moon, though more often and more widely visible than eclipses of the Sun, do not offer by any means the same variety of interesting or striking phenomena to the mere star-gazer, and it was long thought that they were in a certain sense of no use to science. Now, however, astronomers are inclined to utilise them for determining the diameter of the Moon by noting occultations[116] of stars by the Moon, the duration of a star's invisibility behind an eclipsed Moon being a measure of the lunar diameter when such an observation is properly transformed and "reduced." Observations of the heat radiated (or rather reflected) by an eclipsed Moon have also been made with the interesting result of showing that during an eclipse the Moon's power to reflect solar heat to the Earth sensibly declines.

The duration of an eclipse of the Moon is dependent on its magnitude. Where the eclipse is total the darkness, or what counts for such, may last for nearly 4 hours, though this is an extreme limit rarely attained. An eclipse of from 6 to 12 digits (to use the old-fashioned nomenclature which has been already explained) will continue from 2?to 3?hours. An eclipse of 3 to 6 digits will last 2 or 3 hours, and a smaller eclipse only 1 or 2 hours. The visual observations to be made in connection with partial or total eclipses of the Moon chiefly relate to the appearances presented by our satellite when immersed in the Earth's shadow. On such occasions, as has been already stated, it frequently happens that the Moon does not wholly disappear, but may be detected either with a telescope or even without one. It may exhibit either a dull grey appearance, or more commonly a pinkish-red hue to which the designation "coppery" is generally applied. Perhaps the most remarkable instance of this was the eclipse of March 19, 1848.

Mr. Forster who observed the phenomenon at Bruges thus describes[117] what he saw:--"I wish to call your attention to the fact which I have clearly ascertained, that during the whole of the late eclipse of March 19 the shaded surface presented a luminosity quite unusual, probably about three times the intensity of the mean illumination of the eclipsed lunar disc. The light was of a deep red colour. During the totality of the eclipse the light and dark places on the face of the Moon could be almost as well made out as on an ordinary dull moonlight night, and the deep red colour where the sky was clearer was very remarkable from the contrasted whiteness of the stars. My observations were made with different telescopes, but all presented the same appearance, and the remarkable luminosity struck everyone. The British Consul at Ghent, who did not know there was an eclipse, wrote to me for an explanation of the blood-red colour[118] of the Moon at 9 o'clock."

In striking contrast to this stands the total eclipse of Oct. 4, 1884, which is described by Mr. E. Stone[119] as "much the darkest that I have ever seen, and just before the instant of totality it appeared as if the Moon's surface would be invisible to the naked eye during totality; but such was not the case, for with the last appearance of the bright reflected sunlight there appeared a dim circle of light around the Moon's disc, and the whole surface became faintly visible, and continued so until the end of totality."

A total eclipse of the Moon which happened on January 28, 1888, was observed in many places under exceptionally favourable circumstances as regards weather. The familiar copper colour is spoken of by many observers. The Rev. S. Perry makes mention[120] of patches of colour even as bright as "brick red, almost orange in the brighter parts," and this, 20 minutes before the total phase began. Mr. Perry conducted on this occasion spectroscopic observations for the first time on an eclipsed Moon, but no special results were obtained.

Various explanations have been offered for these diversities of appearance. Undoubtedly they depend upon differences in the condition of the Earth's atmosphere, such as the unusual presence or unusual absence of aqueous vapour; but it cannot be said that the laws which control these diversities are by any means capable of being plainly enunciated, notwithstanding that the explanation generally in vogue dates from as far back as the time of Kepler. He suggested that the coppery hue was a result of the refraction of the Earth's atmosphere which had the effect of bending the solar rays passing through it, so that they impinged upon the Moon even when the Earth was actually interposed between the Sun and the Moon. That the outstanding rays which became visible are red may be considered due to the fact that the blue rays are absorbed in passing through the terrestrial atmosphere, just as both the eastern and western skies are frequently seen to assume a ruddy hue when illuminated in the morning or evening by the solar rays at or near sunrise or sunset.

Owing to the variable meteorological condition of our atmosphere, the actual quantity of light transmitted through it is liable to considerable fluctuations, and no wonder therefore that variations occur in the appearances presented by the Moon during her immersion in the Earth's shadow.

It has been suggested that if the portion of the Earth's atmosphere through which the Sun's rays have to pass is tolerably free from aqueous vapour, the red rays will be almost wholly absorbed, but not the blue rays; and the resulting illumination will either only render the Moon's surface visible with a greyish blue tinge, or not visible at all. This will yield the "black eclipse"--to recall the phrase quoted elsewhere. If, on the other hand, the region of the Earth's atmosphere through which the Sun's rays pass be highly saturated, it will be the blue rays which suffer absorption, whilst the red rays will be transmitted and will impart a ruddy hue to the Moon. Finally, if the Earth's atmosphere is in a different condition in different places, saturated in some parts and not in others, a piebald sort of effect will be the result, and some portions of the Moon's disc will be invisible, whilst others will be more or less illuminated. Further illustrations of all these three alternatives will be found amongst the eclipses of the Moon recorded in the chapter[121] devoted to historical matters.

A few instances are on record of a curious spectacle connected with eclipses of the Moon which must have a word of mention. I refer to the simultaneous visibility of the Sun and the Moon above the horizon, the Moon at the time being eclipsed. At the first blush of the thing this would seem to be an impossibility, remembering that it is a cardinal principle of eclipses, both of the Sun and of the Moon, that the three bodies must be in the same straight line in order to constitute an eclipse. The anomalous spectacle just referred to is simply the result of the refraction exercised by the Earth's atmosphere. The setting Sun which has actually set has apparently not done so, but is displaced upwards by refraction. On the other hand, the rising Moon which has not actually risen is displaced upwards by refraction and so becomes, as it were, prematurely visible. In other words, refraction retards the apparent setting of one body, the Sun, and accelerates the apparent rising of the other body, the Moon. The effect of these two displacements will be to bring the two bodies closer by more than 1?of a great circle than they really are, this being the conjoint amount of the double displacements due to refraction.

Amateur observers of eclipses of the Moon will find some pleasure, and profit as well, in having before them on the occasion of an eclipse a picture of the Moon's surface in diagrammatic form with a few of the principal mountains marked thereon; and then watching from time to time (say by quarters of an hour) the successive encroachments of the Earth's shadow on the Moon's surface and the gradual covering up of the larger mountains as the shadow moves forward. The curved lines represent the gradual progress of the shadow during the eclipse named. This diagram, ignoring the curved lines actually marked on it, may be used over and over again for any number of eclipses, simply noting from the Nautical Almanac or other suitable ephemerides the points on the Moon's disc at which the shadow first touches the disc as it comes on, and last touches the disc as it goes off. The Almanac indicates these points by stating that the eclipse begins, or ends, as the case may be, at a point which is so many degrees from the N. point of the Moon measured round the Moon's circumference by the E. or by the W. as the case may be.

One other point and we have disposed of eclipses of the Moon. The shadow which we see creeping over the Moon during an eclipse is, as we know, the shadow cast by the Earth. If we notice it attentively we shall see that its outline is curved, and that it is in fact a complete segment of a circle. Moreover that the circularity of this shadow is maintained from first to last so far as we are able to follow it. What is this, then, but a proof of the rotundity of the earth? This shape of the Earth's shadow on the Moon during a lunar eclipse was suggested as a proof of the rotundity of the Earth by two old Greek astronomers, Manilius and Cleomedes, who lived about 2000 years ago, and is one more illustration of the great powers of observation and the general acuteness of the natural philosophers of antiquity.

FOOTNOTES:

[Footnote 112: The time occupied by the Moon in passing through the penumbra, before and after a lunar eclipse, will generally run to about an hour for each passage. It will occasionally happen that the Moon gets immersed in a penumbra but escapes the dark shadow. Such an event will not be announced in the almanacs under the head of "Eclipses."]

[Footnote 113: See p. 197 (post).]

[Footnote 114: The shadow is spoken of as being in the form of a cone because it is necessarily such on account of the light-giving disc of the Sun being so enormously larger in diameter than the light-receiving sphere of the Moon. This idea can be pursued by any reader with the aid of a lamp enclosed in a glass globe and an opaque sphere such as a cricket ball.]

[Footnote 115: See p. 19 (ante).]

[Footnote 116: As to occultations see chap. xxi. (post).]

[Footnote 117: Month. Not., R.A.S., vol. viii. p. 132. March, 1848.]

[Footnote 118: A very striking chromolithograph of the lunar eclipses of Oct. 4, 1884, and Jan. 28, 1888, showing the contrast of--(1) an almost invisible grey Moon, and (2) a reddish-pink Moon, will be found in the German astronomical monthly, Sirius, vol. xxi. p. 241. Nov. 1888.]

[Footnote 119: Month. Not., R.A.S., vol. xlv. p. 35.]

[Footnote 120: Month. Not., R.A.S., vol. xlviii. p. 227. March 1888.]

[Footnote 121: p. 197 (post).]

CHAPTER XVI.

ECLIPSES OF THE MOON MENTIONED IN HISTORY.

We saw in a previous chapter that we owe to the Chinese the first record of an eclipse of the Sun. It must now be stated that the same remark applies to the first recorded eclipse of the Moon, and Prof. S. Russell is again our authority. He refers to a book called the Chou-Shu or book of the Chou Dynasty, said to have been found in 280 A.D. in the tomb of an Emperor who lived many centuries previously. In this book it is stated that in the 35th year of Wen-Wang on the day Ping-Tzu there was an eclipse of the Moon. Russell finds that this event may be assigned to January 29, 1136 B.C., and that the eclipse was total.

Next after this Chinese eclipse, in point of time, come several eclipses recorded by Ptolemy, on the authority of records collected or examined by himself. The three earliest of these came from Chald鎵 n sources.

The first of these eclipses was observed at Babylon, in the 27th year of the era of Nabonassar, the 1st of the reign of Mardokempadius, on the 29th of the Egyptian month Thoth, answering to March 19, 721 B.C. The eclipse began before moonrise, and the middle of the totality appears to have occurred at 9h.?0m. p.m. The other two eclipses, also observed at Babylon, occurred on March 8, 720 B.C., and September 1, in the same year, respectively.

Three other lunar eclipses, recorded by Ptolemy, assisted Sir I. Newton in fixing the Terminus a quo from which the "70 weeks" of years were to run which the prophet Daniel[122] predicted were to elapse before the death of Christ. This Terminus a quo dates from the Restoration of the Jews under Artaxerxes, 457 B.C. The three eclipses which Newton made use of were those of July 16, 523, November 19, 502, and April 25, 491 B.C.

Aristophanes, in "The Clouds" (lines 561-66), makes an allusion to which has been supposed (but probably without adequate warrant, in Spanheim's opinion), to refer to an eclipse of the Moon. The eclipse, October 9, 425 B.C., has, moreover, been suggested as that referred to, but the whole idea seems to me too shadowy.

An eclipse of the Moon took place in the 4th year of the 91st Olympiad, answering to August 27, 413 B.C., which produced very disastrous consequences to an Athenian army, owing to the ignorance and incapacity of Nicias, the commander. The army was in Sicily, confronted by a Syracusan army, and having failed, more or less, and sickness having broken out, it was decided that the Athenians should embark and quit the island. Plutarch, in his

Life of Nicias, says:--"Everything accordingly was prepared for embarkation, and the enemy paid no attention to these movements, because they did not expect them. But in the night there happened an eclipse of the Moon, at which Nicias and all the rest were struck with a great panic, either through ignorance or superstition. As for an eclipse of the Sun, which happens at the Conjunction, even the common people had some idea of its being caused by the interposition of the Moon; but they could not easily form a conception, by the interposition of what body the Moon, when at the full, should suddenly lose her light, and assume such a variety of colours. They looked upon it therefore as a strange and preternatural phenomenon, a sign by which the gods announced some great calamity." And the calamity came to pass, but only indirectly was it caused by the Moon!

Plutarch and Pliny both mention that eleven days before the victory of Alexander over Darius, at Arbela in Assyria, there was an eclipse of the Moon. Plutarch's words (Life of Alexander) are, that "there happened an eclipse of the Moon, about the beginning of the festival of the great mysteries at Athens. The eleventh night after that eclipse, the two armies being in view of each other, Darius kept his men under arms, and took a general review of his troops by torch-light." This seems to have led to a great deal of disorderly tumult in the Assyrian camp, a fact which was noticed by Alexander. Several of his friends urged him to make a night attack on the enemy's camp, but he preferred that his Macedonians should have a good night's rest, and it was then that he uttered the celebrated answer, "I will not steal a victory." Plutarch enters upon some rather interesting moral reflections connected with this answer, but which of course are foreign to the subject of this volume. This eclipse happened on September 20, 331 B.C., and was total, the middle of the eclipse being about 8.15 p.m. It follows therefore, that the celebrated battle of Arbela was fought on October 1, 331 B.C.

In 219 B.C. an eclipse of the Moon was seen in Mysia, according to Polybius.[123] The date of September 1 has been assigned for this eclipse which is said to have so greatly alarmed some Gaulish Mercenary troops in the service of Attalus, King of Pergamos, that he had to get rid of them as soon as he could--make terms with them to go home.

On the eve of the battle of Pydna when Perseus, King of Macedonia, was conquered by Paulus Paulus there happened an eclipse of the Moon. Plutarch in his Life of Paulus speaking of his army having settled down in a camp, says:--"When they had supped and were thinking of nothing but going to rest, on a sudden the Moon, which was then at full and very high, began to be darkened, and after changing into various colours, was at last totally eclipsed. The Romans, according to their custom, made a great noise by striking upon vessels of brass and held up lighted faggots and torches in the air in order to recall her light; but the Macedonians did no such thing; horror and astonishment seized their whole camp, and a whisper passed among the multitude that this appearance portended the fall of the king. As for Paulus he was not entirely unacquainted with this matter; he had heard of the ecliptic inequalities which bring the Moon at certain periods under the shadow of the Earth and darken her till she has passed that quarter of obscurity and receives light from the Sun again. Nevertheless, as he was wont to ascribe most events to the Deity, was a religious observer of sacrifices and of the art of divination, he offered up to the Moon 11 heifers as soon as he saw her regain her former lustre. At break of day he also sacrificed oxen to Hercules to the number of 20 without any auspicious sign, but in the twenty-first the desired tokens appeared and he announced victory to his troops, provided they stood upon the defensive."

The astronomical knowledge ascribed in this account to Paulus Paulus constitutes a very interesting feature in this record because the Romans though they were good at most things, were by no means adepts at the science of Astronomy. Livy[124] tells us that Sulpicius Gallus, one of the Roman tribunes, foretold this eclipse, first to the Consul and then, with his leave, to the army, whereby that terror which eclipses were wont to breed in ignorant minds was entirely taken off and the soldiers more and more disposed to confide in officers of so great wisdom and of such general knowledge. This eclipse is often identified with that of June 21, 168 B.C., but Johnson gives reasons why this cannot be the case and that the eclipse in question was that which happened on the night of June 10-11, 167 B.C., and commenced about midnight, whereas the eclipse of 168 B.C. was nearly over when the Moon was above the horizon at Rome. Stockwell, however, fixes on the eclipse of September 3, 172 B.C. as that which was connected with the Battle of Pydna.

Josephus[125] speaking of the barbarous acts of Herod, says:--"And that very night there was an eclipse of the Moon." There has been some controversy respecting the identification of this eclipse (the only one mentioned by Josephus) which also is associated with Herod's last illness, it not having been easy to reconcile some discordant chronological statements connected with the length of Herod's reign and the date when he began to reign. On the whole, probably, we shall be safe in saying that the reference is to the eclipse of March 13, 4 B.C. This was a partial eclipse to the extent of less than half the Moon's diameter, a defalcation of light sufficient, however, to attract public notice even at 3 a.m., seeing that no doubt, even at that hour, the streets of Jerusalem were in a state of turmoil owing to the burning alive by Herod of some seditious Rabbis.

It should be stated, however, that Hind assigns the account by Josephus to the eclipse which occurred on January 9, 1 B.C. On this occasion the Moon passed nearly centrally through the Earth's shadow soon after midnight, emerging at 2.57 a.m. on the early morning of January 10, local Mean Time at Jerusalem.

Tacitus[126] mentions an eclipse of the Moon as having happened soon after the death of Augustus. This has been identified with the eclipse of September 27, A.D. 14. Tacitus says:--"The Moon in the midst of a clear sky became suddenly eclipsed; the soldiers who were ignorant of the cause took this for an omen referring to their present adventures: to their labours they compared the eclipse of the planet, and prophesied 'that if to the distressed goddess should be restored her wonted brightness and splendour, equally successful would be the issue of their struggle.' Hence they made a loud noise, by ringing upon brazen metal, and by blowing trumpets and cornets; as she appeared brighter or darker they exulted or lamented."

There was an eclipse of the Moon on the generally recorded date of the Crucifixion of our Lord, April 3, A.D. 33. Hind found that our satellite emerged from the Earth's dark shadow about a quarter of an hour before she rose at Jerusalem (6h.?6m.m), but the penumbra continued upon her disc for an hour afterwards.

On Jan. 1, A.D. 47, a total eclipse of the Moon was seen at Rome, and on the same night an island rose up in the Aegean Sea.

The total eclipse of Feb. 22, A.D. 72, noted by Pliny,[127] is the first in which it is recorded that Sun and Moon were both visible at the same time, the eclipse occurring when the Sun was rising and the Moon setting.

Trithenius speaks of an eclipse of the Moon observed in the time of Merovos. Johnson identifies it with the eclipse of Sept. 15, 452 A.D. It was from Merovos that the line of French kings known as Merovingians received their name.

On April 16, A.D. 683, according to Anastasius the Papal historian, the Moon for nearly the whole night exhibited a blood-red appearance, and did not emerge from obscurity till cockcrowing.

In A.D. 690 an eclipse of the Moon was observed in Wales. We are told[128] that "the Moon was turned to the colour of blood." This would seem to be the first eclipse of the Moon recorded in Britain.

The Anglo-Saxon Chronicle tells us that in A.D. 734 "the Moon was as if it had been sprinkled with blood, and Archbishop Tatwine and Beda died and Ecgberht was hallowed bishop." The intended inference apparently is that the Moon had something to do with the deaths of the two ecclesiastics, but this theory will not hold water. Beda, it may be remarked, is the correct name of the man generally known to us as the "Venerable Bede." It is evident that from the description of the Moon it exhibited on that occasion the well-known coppery hue which is a recognised feature of many total eclipses of our satellite. This eclipse occurred on January 24, beginning at about 1 a.m.

On the night of January 23, A.D. 753, "the Moon was covered with a horrid black shield." This is the record of an eclipse. It occurred at about midnight, and apparently we are entitled to infer that on this occasion the Moon disappeared altogether, instead of being discoverable during the total phase by exhibiting a coppery hue.

In A.D. 755 [or 756 in orig.], on November 23, there happened an exceedingly interesting event which stands, I think, without a precedent in the annals of science--an eclipse of the Moon contemporaneous with an occultation of a planet by the Moon. This singular combination is thus described in the annals of Roger de Hoveden[129]:--"On the 8th day before the Calends of December the Moon on her 15th day being about her full, appeared to be covered with the colour of blood, and then the darkness decreasing she returned to her usual brightness; but, in a wondrous manner, a bright star followed the Moon, and passing across her, preceded her when shining, at the same distance which it had followed her before she was darkened." The details here given are not astronomically quite correct, but let that pass; the writer's intention is fairly clear. Calculation shows that the eclipse occurred on November 23, and that the planet, which was Jupiter, was concealed in the evening by the Moon for about an hour from 7h.?0m. to 8h.?0m.m, the immersion taking place about the end of the total phase. This is the first occultation of a star or planet by the Moon observed and recorded in England.

Under the year 795 the Anglo-Saxon Chronicle says:--"In this year the Moon was eclipsed between cockcrowing and dawn on the 5th of the Calends of April; and Eardwalf succeeded to the kingdom of the Northumbrians on the 2nd of the Ides of May." This signifies that the eclipse happened on March 28 between 3h. and 6h. in the morning, the method of dividing the hours of night into equal portions of three hours each being still in use. There was no eclipse in 795 on the date in question but there was one in 796, so we may suppose an error in the year. This assumed, Johnson found that the eclipse began at about 4h..m., was total for nearly an hour, and ended at about 7 絵., so that the Moon set eclipsed. But the above assumption is dispensed with by Lynn who substitutes one of his own.[130] For "5th of the Calends" he reads "5th of the Ides," which means April 9; and on that day in 795 he says there was an eclipse of the Moon, but I have not found any other record of it.

In the year A.D. 800, according to the Anglo-Saxon Chronicle, "the Moon was eclipsed at the 2nd hour of the night (8h.m) on the 17th day of the Calends of February." Johnson finds that there was an eclipse of the Moon on Jan. 15. The middle of the eclipse occurred at 8h.?4m., 9/10ths of the Moon's upper limb having been obscured.

Under the date of 806 the Anglo-Saxon Chronicle says:--"This year was the Moon eclipsed on the Kalends [1st] of September; and Eardwulf, King of the Northumbrians, was driven from his kingdom, and Eanberht, Bishop of Hexham, died." This eclipse was total, the totality lasting from 9h.?7m. to 10h.?9m.m

On Feb. 15, 817, according to the Annales Fuldenses, an eclipse of the Moon was observed in the early evening at Paris, and on the same night a Comet was seen. This Comet is described by another authority as a "monstrous" one and as being in Sagittarius on Feb. 5. The Chinese date it for Feb. 17, and place it near the stars [Greek: alpha] and [Greek: gamma] Tauri.

In 828 two lunar eclipses were seen in Europe, the first on July 1 very early in the morning, and the second on the morning of Christmas Day. The Anglo-Saxon Chronicle thus speaks of the second eclipse:--"In this year the Moon was eclipsed on Mid-winter's Mass-night, and the same year King Ecgbryht subdued the kingdom of the Mercians and all that was South of the Humber." The totality occurred after midnight. There is some confusion in the year of this eclipse, the Chronicle giving it as 827, whilst calculation shows that it must have been 828. Lynn defines "Mid-winter's Mass-night" as Christmas Eve.

Under the date of 904 the Anglo-Saxon Chronicle says:--"In this year the Moon was eclipsed." There were two total eclipses of the Moon this year, one on May 31, and the other on Nov. 25, and it does not appear which one is referred to in the Chronicle cited. Another writer, Cedrenus, speaks of a great eclipse of the Moon this year which he says foretold the death of a kinsman of the Emperor.

On October 6, 1009, there was a total eclipse of the Moon which presumably is referred to in the statement that "this year the Moon was changed into blood."

On Nov. 8, 1044, there was a large partial eclipse in the morning. Raoul Glaber[131] (a French chronicler who died about 1050) comments upon it thus:--"In what manner it happened, whether a prodigy brought to pass by the Deity or by the intervention of some heavenly body, remains known to the author of knowledge. For the Moon herself became like dark blood, only getting clear of it a little before the dawn." Truly those times were the "Dark Ages" in which ignorance and folly were rampant, seeing that more than 1000 years previously the Greeks knew all about the causes of eclipses.

Under 1078 the Anglo-Saxon Chronicle says:--"In this year the Moon was eclipsed 3 nights before Candlemas, and 荥 elwig, the 'world-wide' Abbot of

Evesham, died on St. Juliana's Mass-day [Feb. 16]; and in this year was the dry summer, and wildfire came in many Shires and burned many towns." Johnson found that a total eclipse of the Moon happened in the early evening of Jan. 30.

On May 5, 1110, in the reign of Henry I., there occurred a total eclipse of the Moon during which, says the Anglo-Saxon Chronicle, "the Moon appeared in the evening brightly shining and afterwards by little and little its light waned, so that as soon as it was night it was so completely quenched that neither light nor orb nor anything at all of it was seen. And so it continued very near until day, and then appeared full and brightly shining. It was on this same day a fortnight old. All the night the air was very clear, and the stars over all the heaven were brightly shining. And the tree-fruits on that night were sorely nipt." The totality occurred before mid-night. It is evident that this was an instance of a "black" eclipse when the Moon becomes quite invisible instead of shining with the familiar coppery hue.

In 1117 there were two total eclipses, the first on June 16, and the second on December 10. The latter is thus referred to in the Anglo-Saxon Chronicle:-- "In the night of the 3rd of the Ides of December the Moon was far in [during a long time of] the night as if it were all bloody, and afterwards eclipsed." The totality commenced at 11.36 p.m.

It is recorded by Matthew Paris[132] in connection with the death of Henry I. that "the Moon also was eclipsed the same year on the 29th of July" [1135]. These words seem to indicate a total eclipse of the Moon. Johnson gives the date as Dec. 22, 1135. If this is correct the text of the Chronicle must be corrupt. The whole eclipse was not visible in England, the Moon setting before the middle of the eclipse. Stephen had been crowned king the same day, namely Dec. 22.

On June 30, 1349, there was a total eclipse of the Moon visible at London to which some interest attaches. Archdeacon Churton[133] connects it with the following incident:--"The worthy Abp. Bradwardine, who nourished in the reign of the Norman Edwards, and died A.D. 1349, tells a story of a witch who was attempting to impose on the simple people of the time. It was a fine summer's night, and the Moon was suddenly eclipsed. 'Make me good amends,' said she, 'for old wrongs, or I will bid the Sun also to withdraw his light from you.' Bradwardine, who had studied the Arabian astronomers, was more than a match for this simple trick, without calling in the aid of the Saxon law. 'Tell me,' he said, 'at what time you will do this, and we will believe you; or if you will not tell me, I will tell you when the Sun or the Moon will next be darkened, in what part of their orb the darkness will begin, how far it will spread, and how long it will continue.'"

An eclipse of the Moon which happened when Columbus was at the Island of Jamaica proved of great service to him when he was in difficulties owing to the want of food supplies which the inhabitants refused to afford. The eclipse was a total one, and so far as the description goes the eclipses of April 2, 1493, and March 1, 1504, both respond to the recorded circumstances: both were total and both occurred soon after sunset. But, inasmuch as in the life of Columbus written by his son the incident is placed nearly at the end of the work, there can be no doubt that it is the later of the above eclipses which was the one in question. The story is very graphically told by Sir A. Helps[134] in the words following:--

"The Indians refused to minister to their wants any longer; and famine was imminent. But just at this last extremity, the admiral, ever fertile in devices, bethought him of an expedient for re-establishing his influence over the Indians. His astronomical knowledge told him that on a certain night an eclipse of the Moon would take place. One would think that people living in the open air must be accustomed to see such eclipses sufficiently often not to be particularly astonished at them. But Columbus judged--and as the event proved, judged rightly--that by predicting the eclipse he would gain a reputation as a prophet, and command the respect and the obedience due to a person invested with supernatural powers. He assembled the caciques of the neighbouring tribes. Then, by means of an interpreter, he reproached them with refusing to continue to supply provisions to the Spaniards. 'The God who protects me,' he said, 'will punish you. You know what has happened to those of my followers who have rebelled against me; and the dangers which they encountered in their attempt to cross Haiti, while those who went at my command made the passage without difficulty. Soon, too, shall the divine vengeance fall on you; this very night shall the Moon change her colour and lose her light, in testimony of the evils which shall be sent upon you from the skies.'

"The night was fine: the moon shone down in full brilliancy. But at the appointed time the predicted phenomenon took place, and the wild howls of the savages proclaimed their abject terror. They came in a body to Columbus and implored his intercession. They promised to let him want for nothing if only he would avert this judgment. As an earnest of their sincerity they collected hastily a quantity of food and offered it at his feet. At first, diplomatically hesitating, Columbus presently affected to be softened by their entreaties. He consented to intercede for them; and, retiring to his cabin, performed, as they supposed, some mystic rite which should deliver them from the threatened punishment. Soon the terrible shadow passed away from the face of the moon, and the gratitude of the savages was as deep as their previous terror. But being blended with much awe, it was not so evanescent as gratitude often is; and henceforth there was no failure in the regular supply of provisions to the castaways."

Tycho Brahe observed a lunar eclipse on July 7, 1590. He writes:--"In the morning about 3 縱. the Moon began to be eclipsed: in this eclipse it is notable that both luminaries were at the same time above the horizon; a like case which Pliny cites. For the centre of the Sun emerged when the Moon was 2?elevated above the Western horizon, and when her centre was setting, the centre of the Sun was elevated nearly 2?"[135]

On August 16, 1598, there occurred a total eclipse of the Moon, observed by Kepler,[136] in which during totality a part of the Moon was visible and the rest invisible. He says, that while one-half of the disc was seen with great difficulty the other half was discernible by a deep red light of such brilliancy that at first he was doubtful whether our satellite was immersed in the Earth's shadow at all. This is an instance of the simultaneous operation of those causes (whatever they may be) which result in a totally-eclipsed Moon being sometimes wholly invisible and sometimes entirely visible as a copper-coloured disc.

An eclipse of the Moon which happened on the morning of July 6, 1610, may be mentioned as having been the first to be viewed through a telescope. The eclipse was only a large partial one. The following record of the fact is due to Tycho Brahe.[137] "The beginning of the eclipse of the Moon as observed through the Roman telescope, appeared like a dark thread in contact with the shadow"--a description which cannot be said to be unduly explicit.

In 1620, on June 15, there was a total eclipse of the Moon, when during the total phase "the Moon was seen with great difficulty. It shone, moreover, like the thinnest nebula, far fainter than the Milky Way, without any copper tinge. About the middle of the second hour nothing at all could be seen of the Moon with the naked eye, and through the telescope so doubtfully was anything seen that no one could tell whether the Moon was not something else." It is expressly stated, however, that the sky was quite clear. Kepler also observed this eclipse, and says that the Moon quite disappeared, though stars of the 4th and 5th magnitudes were plainly visible.[138] In this same year 1620, there was on December 9 another total eclipse, when "the Moon altogether disappeared so that nothing could be seen of it, though the stars shone brightly all around: she continued lost and invisible for a quarter of an hour more or less." This observation seems to have been made at Ingolstadt.

Wendelinus mentions the eclipse of April 14, 1623, in connection with the question of the visibility of the Moon when totally eclipsed. He says, "but sometimes it so far retains the light derived from the Sun that you would doubt whether any part of it were eclipsed." This eclipse was observed by Gassendi, and if the above record is correct, it is the more remarkable seeing that the eclipse was not total, only 11/12ths of the Moon's diameter being obscured.

On April 25, 1642, on the occasion of a total eclipse, Hevelius[139] noted that the Moon wholly disappeared when immersed in the Earth's shadow. Crabtree is stated by Flamsteed[140] to have observed this eclipse, but he does not plainly state that he lost sight of the Moon. Crabtree or his editor dates this eclipse for April 4; Ferguson for April 15. There appears to be some muddle as between "old style" and "new style." Ferguson professing to be N.S. is evidently wrong. Hevelius gives the double date, 15/25, which is evidently right.

On June 16, 1666, the Moon was seen in Tuscany to rise eclipsed, the Sun not having yet set in the W.

On May 26, 1668, an eclipse of the Moon was in progress in the early morning, when the Sun was seen to rise by members of the Academy of Sciences who were observing the phenomenon at Montmartre near Paris.

On December 23, 1703, the Moon when totally immersed was seen at Avignon showing a ruddy light of such brilliancy that we are told it had the appearance of a transparent body illuminated by a light placed behind. Johnson finds that the total phase took place in the early morning, and lasted from 5h.?6m. to 7h.?2mm.

The lunar eclipse of May 18, 1761, as observed by Wargentin,[141] at Stockholm, furnishes a remarkable instance of the invisibility of the Moon on certain occasions, when completely immersed in the earth's shadow. The total immersion of the Moon took place at 10h.?.m. The part of the margin of the lunar disc which had last entered the shadow was fairly conspicuous for 5 or 6 minutes after the immersion, and to the naked eye exhibited a lustre equal to that of a star of the 2nd magnitude; but at 10h.?2m. this part, as well as the whole of the rest of the Moon's body, "had disappeared so completely, that not the slightest trace of any portion of the lunar disc could be discerned either with the naked eye or with the telescope, although the sky was clear, and the stars in the vicinity of the Moon were distinctly visible in the telescope." After more than half an hour's search, Wargentin at length discovered the whereabouts of the Moon by means of a faint light, which was visible at the Eastern edge of the disc. A few minutes afterwards, some persons of acute vision were able to discern, with the naked eye, a trace of the Moon, looking like a patch of thin vapour, but more than half the disc was still invisible.

An eclipse of the Moon, on March 29, 1801, was observed by Humboldt, on board ship, off the Island of Baru, not far from Cartagena de las Indias, in the Caribbean Sea.[142] He remarks that he was "exceedingly struck with the greater luminous intensity of the Moon's disc under a tropical sky than in my native North." Johnson makes Humboldt to refer to the greater clearness of the "reddened disc," but these words do not appear either in the German or in the English version.

A total eclipse of the Moon occurred on June 10, 1816. As observed by Beer and M 銃 ler and others, the Moon completely disappeared. The summer of 1816, be it remembered, was very wet, and probably this had something to do with the Moon's invisibility at the eclipse in question.

On October 13, 1837, there happened a total eclipse of the Moon, of which Sir J. Herschel and Admiral W.燃. Smyth have left us interesting accounts.[143] The changes of tint, both as regards times and places on the Moon's disc, recorded by the latter, are very remarkable. And the tints themselves varied very much inter se: The Admiral speaks of "copper," "sea-green," "neutral tint," and "silvery," as hues visible in one part of the Moon or another, and at one time or another.

FOOTNOTES:

[Footnote 122: Dan. ix. 24.]

[Footnote 123: Histories, Book v., chap. lxxviii.]

[Footnote 124: Hist. Rom., Lib. xliv., cap. 37.]

[Footnote 125: Antiq., Lib. xvii., cap. 6, sec. 4.]

[Footnote 126: Annales, Lib. i., cap. 28.]

[Footnote 127: Nat. Hist., Lib. ii., cap. 3.]

[Footnote 128: Annales Cambri? Rolls ed., p. 8.]

[Footnote 129: Annales, Rogerus de Hoveden, Bohn's ed., p. 5.]

[Footnote 130: Observatory, vol. xv. p. 224. May 1892.]

[Footnote 131: Historiarum sui Temporis, Lib. v., cap. 3.]

[Footnote 132: Chronica Majora, Rolls ed., edited by the Rev. H.Luard, vol. ii. p. 161. Another version of this work is in circulation under the name of Rogerus de Wendover, Flores Historiarum. The passage here quoted appears in vol. i. p. 482, Bohn's ed.]

[Footnote 133: History of the Early English Church, 1870 ed., p. 271.]

[Footnote 134: Life of Columbus, p. 247.]

[Footnote 135: Historia Coelestis, vol. i. p. xci.]

[Footnote 136: Astronomi?Pars Optica, p. 276; Opera Omnia, vol. ii. p. 302; Frisch's edition.]

[Footnote 137: Historia Coelestis, vol. ii. p. 921.]

[Footnote 138: Epitomes Astronomi? p. 825; Opera Omnia, vol. vi. p. 482; Frisch's edition.]

[Footnote 139: Selenographia, p. 117.]

[Footnote 140: Historia Coelestis, vol. i. p. 4.]

[Footnote 141: Phil. Trans., vol. lii. p. 210. 1762.]

[Footnote 142: Cosmos. Trans. Sabine, vol. iii. p. 356; vol. iv. p. 483. Bohn's ed.]

[Footnote 143: Cycle of Celest. Obj., vol. i. p. 144; transcribed in G.燜. Chambers's Handbook of Astronomy, vol. i. p. 329.]

CHAPTER XVII.

CATALOGUES OF ECLIPSES: AND THEIR CALCULATION.

This must of necessity be a brief chapter, so far as mere lines of text are concerned, but it will not on that account be unimportant. It will be evident to the reader that many more eclipses of interest have happened, and will happen, than it has been possible to speak of in these pages. Accordingly, as it is one of the main objects of this series of volumes to create a thirst for knowledge, to be satisfied by the study of other and bigger volumes, it will be desirable to furnish a list of some of the various books and publications, in which eclipses will be found catalogued or described in detail, so that readers desirous of pursuing the matter further, may possess facilities for doing so.

By far the most complete and comprehensive catalogue of solar eclipses is that prepared some years ago by an Austrian astronomer, the late Theodore Von Oppolzer of Vienna, and published under the title of Canon der Finsternisse, in the Memoirs of the Imperial Academy of Sciences.[144] This work supplies approximate calculations of about 8000 eclipses of the Sun, for a period of more than 3000 years, from November 10, 1207 B.C. (Julian Calendar), to November 17, 2161 A.D. (Gregorian Calendar). There are appended 160 charts, of all the principal eclipses; but as the charts only exhibit the beginnings, middles, and ends of the eclipses dealt with, they are frequently misleading, because the intermediate lines of path are, in many cases, more or less considerably curved.

Another very important and comprehensive catalogue of eclipses, solar and lunar together, will be found in the well-known French work, L'Art de vie fier les Dates,[145] compiled by a member of the religious order of St. Maur. One volume of this famous work contains eclipses from the year 1001 B.C. to the Christian Era, whilst another volume gives a similar catalogue from the year 1 A.D. to 2000 A.D. The other volumes deal with chronological matters only. Although not strictly a work of extreme astronomical exactness, yet L'Art de vie fier les Dates stands unrivalled as a record not only to subserve the purpose indicated by its title, but of the bare facts of the eclipses which have happened during the period of 3000 years stated above.

There has not been much done in England in the way of publishing eclipse records or tables, past or future, but in the British Almanac and Companion for 1832 there is given a catalogue, which was useful in its day, of eclipses, then future from 1832 to 1900, omitting, however, solar eclipses hardly visible to any inhabited portion of the Earth, and lunar eclipses where the part of the Moon's diameter obscured was less than 1/12th.

In by-gone days several attempts were made to gather together in a tabular or paragraph form the details of eclipses which had happened, and some of these have been important sources of information for the guidance of us moderns. Foremost amongst these efforts must be named the Almagestum Novum of J. Ricciolus.[146] This work contains a catalogue of eclipses observed from 772 B.C. to A.D. 1647, and continued in tables to A.D. 1700. It is prefaced (pp. 286-8) by a long series of quotations from classical authors relating to eclipses, some few of which have already been mentioned in these pages.

Kepler paid much attention to eclipses, and left behind him a large mass of notes and original observations. These will be found chiefly in his Astronomi?Pars Optica, c. vii. ?2, originally published at Frankfurt in 1604. The most convenient and accessible edition of this is to be found in Frisch's reprint of all Kepler's works.[147]

Tycho Brahe also gathered together from various sources many observations of eclipses, and combined them with a number of his own, the whole being published in his Historia Coelestis.[148] Tycho Brahe was a very interesting personage in spite of the fact that he went all astray on the subject of the system of the Universe, and he well deserves, what has been given to him, a book[149] all to himself. It is peculiarly appropriate that I should give him a good word in this little volume on eclipses, because it was the solar eclipse of Aug. 21, 1560, which first seriously led him to take up astronomical pursuits, he being then 14 years of age, and struck with wonder that eclipses could be predicted.

A vast amount of historical and other information respecting eclipses will be found in a book, the latinised name of whose author is Sethus Calvisius. The title of the work is Opus Chronologicum.[150] The historical matter is very much mixed, but the eclipses can be got hold of through the Index, which is very full. P. Gassendi,[151] a well-known astronomer of the 17th century, left behind him observations of many eclipses observed by himself between 1628 and 1655. In a book entitled An Introduction to Universal Geography,[152] one Nicolas Struyck in the middle of the 18th century published a very full array of eclipse observations collected with infinite pains from an endless variety of authors ancient and modern.

In 1757 the well-known James Ferguson reprinted in his Astronomy,[153] but in a very condensed form, all Struyck's eclipses from 721 B.C. to A.D. 1485. Then he carried on his catalogue to 1800 by means of the materials furnished by Ricciolus and L'Art de vie ifier les Dates. Ferguson also invented a machine for illustrating mechanically the circumstances of an eclipse. He called it the "Eclipsareon." A full description is given in his book, mentioned above, but I do not know whether any such instrument is still in existence, or, if so, where it is to be found.

Ferguson apologises[154] for the incompleteness of his eclipse information in the following words:--"I have not cited one half of Ricciolus's list of portentous eclipses, and for the same reason that he declines giving any more of them than what that list contains, namely, that 'tis most disagreeable to dwell any longer on such nonsense, and as much as possible to avoid tiring the reader. The superstition of the ancients may be seen by the few here copied. My author further says that there were treatises written to show against what regions the malevolent effects of any particular eclipse was aimed, and the writers affirmed that the effects of an eclipse of the Sun continued as many years as the eclipse lasted hours, and that of the Moon as many months."

The most comprehensive (indeed almost the only) modern English book on eclipses is the Rev. S.Johnson's,[155] of which frequent use has already been made in these pages. It contains a vast amount of matter put together in a condensed form but the references to authorities are rather defective and deficient. Less comprehensive in one sense yet exceedingly valuable and interesting as a succinct summary of solar eclipse knowledge up to the date of 1896 is Mrs. Todd's excellent little volume[156] which has been several times quoted on previous pages. On various occasions in 1890 and following years Professor J.燦. Stockwell contributed to the American Astronomical Journal a number of papers[157] discussing in a very interesting and exhaustive manner many of the eclipses recorded by the ancient classical authors. These papers should be consulted by all who desire to realise the value of eclipse records in connection with mundane chronology.

The calculation of eclipses is a matter of some interest. It is beyond the scope of the present work to explain even in outline the methods in use, but with the aid of the books mentioned below[158] a reader possessed of the necessary time, mathematical knowledge, and patience, will be able to pursue this matter as far as his inclination may lead him. Johnson has found very useful the tables given in the eighth edition of the Encyclopedia Britannica (Article, "Astronomy") but strange to say these tables do not appear in ninth edition of that famous work.

Lalande[159] has given numerous references to eclipses of the Sun during the 16th, 17th and 18th centuries which may be useful to those who wish to work at the history of eclipses.

FOOTNOTES:

[Footnote 144: Denkschriften der Kaiserlichen Akademie der Wissenschaften, vol. lii. Vienna, 1887.]

[Footnote 145: There are several editions of this work in circulation. The first (published in 1783) was in folio volumes, but the best known edition is in a large number of octavo volumes published in 1818 and following years. The eclipse lists will be found in the 1st volumes of the first and second series respectively. The French astronomer, Pingr? is responsible for them.]

[Footnote 146: Published at Bononia (Bologna) in 1653.]

[Footnote 147: Omnia Opera, vol. ii. pp. 311-16. Edited by Ch. Frisch. 8 vols. 8vo. Frankofurti-a-M., 1857-60.]

[Footnote 148: A collected edition of Tycho Brahe's works, edited by "Lucius Barettus," was published at August?Vindilicorum (Augsburg) in 1666. Lucius Barettus is an anagram for the real name Albertus Curtius.]

[Footnote 149: Dreyer, Tycho Brahe: a Picture of Scientific Life and Work in the Sixteenth Century.]

[Footnote 150: Opus Chronologicum. Francofurti ad Moenum, 1650.]

[Footnote 151: Astronomica, vol. iv. Lugduni, 1657.]

[Footnote 152: Inleiding tot de Algemeene Geographie. Amsterdam, 1740.]

[Footnote 153: Astronomy Explained upon Sir Isaac Newton's Principles. 2nd ed. 4to, pp. 167-79. London, 1757.]

[Footnote 154: Astronomy, p. 178.]

[Footnote 155: Historical and Future Eclipses. 2nd Ed., 1896.]

[Footnote 156: Total Eclipses of the Sun. Boston, U.S., 1894.]

[Footnote 157: Astronomical Journal, vol. x. pp. 25, 185; vol. xi. pp. 5, 28, 57; vol. xii. p. 121; vol. xiii. p. 73; vol. xv. p. 73; vol. xvi. pp. 89, 175.]

[Footnote 158: J. Ferguson. Op. cit.; W.燚. Snooke, Brief Astronomical Tables for the Expeditious Calculation of Eclipses, 8vo. Lond. 1852.]

CHAPTER XVIII.

STRANGE ECLIPSE CUSTOMS.

I had intended heading this chapter "Eclipse Customs amongst Barbarous Nations," but in these days it is dangerous to talk of barbarians or to speak one's mind on points of social etiquette so I have thought it well to tone down the original title, otherwise I should have the partisans of the "Heathen Chinee" holding me up to scorn as a reviler of the brethren.

Did space permit a very interesting record might be furnished of eclipse customs in foreign parts.

An eclipse happened during Lord Macartney's embassy to China[160] which kept the Emperor and his Mandarins for a whole day devoutly praying the gods that the Moon might not be eaten up by the great dragon which was hovering about her. The next day a pantomime was performed, exhibiting the battle of the dragon and the Moon, and in which two or three hundred priests, bearing lanterns at the end of long sticks, dancing and capering about, sometimes over the plain, and then over chairs and tables, bore no mean part.

Professor Russell, who is quoted elsewhere in this work with respect to Chinese eclipses, makes the following remarks in regard to what happens now in China when eclipses occur:--"It will be interesting here to note that, even at present, by Imperial command, special rites are performed during solar and lunar eclipses. A president from each of the six boards, with two inferior officials, dressed in their official clothes, proceed to the T'ai-Ch'ang-Ssu. When the eclipse begins they change their robes for common garments made of plain black material, and kneeling down, burn incense. The president then beats one stroke on a gong, and the ceremony is taken up by all the attendant officials."

A writer in Chambers's Journal[161] in an article entitled "The Hindu view of the late Eclipse," gives an interesting and original account of divers Hindu superstitions and ceremonies which came under his notice in connection with the total eclipse of the Sun of Aug. 18, 1868. He remarks that "European science has as yet produced but little effect upon the minds of the superstitious masses of India. Of the many millions who witnessed the eclipse of the 18th of August last there were comparatively few who did not verily believe that it was caused by the dragon Rahu in his endeavour to swallow up the Lord of Day.... The pious Hindu, before the eclipse comes on, takes a torch, and begins to search his house and carefully removes all cooked food, and all water for drinking purposes. Such food and water, by the eclipse, incur Grahana seshah, that is, uncleanness, and are rendered unfit for use. Some, with less scruples of conscience, declare that the food may be preserved by placing on it dharba or Kusa grass," and much more to the like effect is duly set out in the interesting article cited.

During the total eclipse of the Sun of Aug. 7, 1869, the following incident is noted[162] to have occurred at a station on the Chilkaht river, in Alaska, North America, frequented by Indians:--

"About the time the Sun was half obscured the chief Koh-Klux and all the Indians had disappeared from around the observing tent; they left off fishing on the river banks; all employments were discontinued; and every soul disappeared; nor was a sound heard throughout the village of 53 houses. The natives had been warned of what would take place, but doubted the prediction. When it did occur they looked upon me as the cause of the Sun's being 'very sick and going to bed.' They were thoroughly alarmed, and overwhelmed with an undefinable dread."

A still more thrilling incident is thus recorded[163] of the eclipse of July 29, 1878, by a witness at Fort Sill, Indian Territory, U.S.:--

"On Monday last we were permitted to see the eclipse of the Sun in a beautiful bright sky. Not a cloud was visible. We had made ample preparation, laying in a stock of smoked glass several days in advance. It was the grandest sight I ever beheld, but it frightened the Indians badly. Some of them threw themselves upon their knees and invoked the Divine blessing; others flung themselves flat on the ground, face downward; others cried and yelled in frantic excitement and terror. Finally one old fellow stepped from the door of his lodge, pistol in hand, and fixing his eyes on the darkened Sun, mumbled a few unintelligible words and raising his arm took direct aim at the luminary, fired off his pistol, and after throwing his arms about his head in a series of extraordinary gesticulations retreated to his own quarters. As it happened, that very instant was the conclusion of totality. The Indians beheld the glorious orb of day once more peep forth, and it was unanimously voted that the timely discharge of that pistol was the only thing that drove away the shadow and saved them from the public inconvenience that would have certainly resulted from the entire extinction of the Sun."

A certain Mr. F. Kerigan, in a book published in 1844, made the following remarks on ancient Jewish ideas respecting eclipses:--

"The Israelites, like their benighted neighbours, esteemed an eclipse of either luminary as a supernatural and inauspicious omen, which filled them with the most gloomy and fearful apprehensions: as may fairly be deduced from the 8th chapter of Ezekiel, v. 15: 'Then he brought me to the door of the Lord's House, which was towards the N.; and, behold there sat women weeping for Tammuz.' Now Tammuz is the name under which Adonis was known in Palestine: he was the favourite of Venus, or Astarte, the principal goddess of the Philistines and Phoenicians. Being killed by a wild boar, the prevailing superstition of the age induced the uninformed multitude to believe that when the Moon was eclipsed, it was in complement to their beloved goddess Venus or Astarte, who, concealed behind the full Moon, sat weeping under a dark veil for the loss of her beloved Tammuz or Adonis."[164]

The African travellers, R. and J. Lander, have given[165] a graphic account of what took place on the occasion of the eclipse of the Moon of Sept. 2, 1830, as witnessed by themselves:--"The earlier part of the evening had been mild, serene, and remarkably pleasant. The Moon had arisen with uncommon lustre, and being at the full, her appearance was extremely delightful. It was the conclusion of the holidays, and many of the people were enjoying the delicious coolness of a serene night, and resting from the laborious exertions of the day; but when the Moon became gradually obscured, fear overcame every one. As the eclipse increased they became more terrified. All ran in great distress to inform their sovereign of the circumstance, for there was not a single cloud to cause so deep a shadow, and they could not comprehend the nature or meaning of an eclipse.... Groups of men were blowing on trumpets, which produced a harsh and discordant sound; some were employed in beating old drums, others again were blowing on bullocks' horns.... The diminished light, when the eclipse was complete, was just sufficient for us to distinguish the various groups of people, and contributed in no small degree to render the scene more imposing. If a European, a stranger to Africa, had been placed on a sudden in the midst of the terror-struck people, he would have imagined himself to be among a legion of demons, holding a revel over a fallen spirit."

FOOTNOTES:

[Footnote 159: Bibliographie Astronomique. Paris, 1803. Indexed at p. 938.]

[Footnote 160: Authentic Account of an Embassy to China, by Sir G. Staunton.]

[Footnote 161: Fourth Series, vol. v. p. 676. October 24, 1868.]

[Footnote 162: Report U.S. Coast Survey, 1869, p. 179.]

[Footnote 163: Letter published in the Philadelphia Inquirer.]

[Footnote 164: A Practical Treatise on Eclipses, p. 2.]

[Footnote 165: Journal of an Expedition to Explore the Niger, vol. i. p. 366.]

CHAPTER XIX.

ECLIPSES IN SHAKESPEARE AND THE POETS.

The sound of these words may be large but facts do not bear out the theory, for eclipses do not appear to have captivated our great poets to anything like the extent that Moon, Stars, and Comets have done.

Shakespeare has a few allusions to eclipses, but they are not of prime importance. In Macbeth we find:--

"And slips of yew Shivered in the Moon's eclipse" --Act iv. sc. 1.

the precise meaning of which is not very obvious. "Shivered" of course means divided into pieces, but the idea intended is obscure.

The next quotation is more comprehensive and reflects more plainly the current of thought prevalent in Shakespeare's day, albeit here again the word "eclipse" will be found to stand without much definite connection with what goes before. However the reader shall judge for himself:--

"As stars with trains of fire and dews of blood, Disasters in the Sun; and the moist star, Upon whose influence Neptune's Empire stands, Was sick almost to doomsday with eclipse." --Hamlet, act i. sc. 1.

In King Lear we seem to come upon something very definitely historical, but I am not able to say what it is. The Earl of Gloster says:--

"These late eclipses in the Sun and Moon portend no good to us."

With this, Edmund, Gloster's son, apparently agrees, for he exclaims:--

"These eclipses do portend these divisions." --Act i. sc. 2.

In Othello, the Moor of Venice himself, in a moment of excitement, says:--

"O, insupportable! O, heavy hour! Methinks it should be now a huge eclipse Of Sun and Moon, and that the affrighted globe Should yawn at alteration." --Act v. sc. 2.

In Anthony and Cleopatra we find Anthony expressing what our forefathers so often thought in connection with astronomical matters:--

"Alack, our terrine Moon is now eclipsed; And it portends alone The fall of Anthony!" --Act iii. sc. 11.

Milton has an allusion to an eclipse of the Sun which possesses a two-fold interest--intrinsic and extrinsic. The former feature will be self-evident when the passage is read. The poet, in describing[166] the faded splendour of the fallen archangel, compares him to the Sun seen under circumstances which have temporarily deprived it of its normal brilliancy and glory:--

"As when the Sun new-risen Looks through the horizontal misty air Shorn of his beams, or, from behind the Moon In dim eclipse, disastrous twilight sheds On half the nations, and with fear of change Perplexes Monarchs."

It has been well said by Dr. Orchard[167] that "this passage affords us an example of the sublimity of Milton's imagination and of his skill in adapting the grandest phenomena of nature to the illustration of his subject."

What I alluded to in saying that extrinsic interest attached to this quotation, is the fact that these lines might have caused the suppression of the poem as a whole. Mrs. Todd puts the matter thus:--"Paradise Lost was begun probably in 1658, although not finished until 1663, nor its thorough revision completed until 1665. The censorship still existed, and Tomkyns (one of the chaplains through whom the Archbishop gave or refused license), although a broader-minded man than many of his day, found this passage especially objectionable. The poem was allowed to see the light only through the interposition of a friend of Milton. Upon such slender chances may hang the life of an incomparable work of art! But it is easy to see that in the turbulent days when Charles the Second had returned to power, after the death of Cromwell, these lines should have been deemed dangerously suggestive, in imputing to monarchs 'perplexity' and 'fear of change.'"

Other allusions to eclipses by Milton will be found as follows:--

Through the air she comes, "Lur'd with the smell of infant blood, to dance With Lapland witches, while the labouring Moon Eclipses at their charms." -- Paradise Lost, Bk. ii. lines 663-6.

"So saying, he dismiss'd them; they with speed Their course through thickest constellation held, Spreading their bane; the blasted stars look'd wan, And planets, planet-struck, real eclipse, Then suffer'd." --Paradise Lost, Bk. x. lines 410-14.

"O dark, dark, dark, amid the blaze of Noon, Irrecoverably dark, total eclipse,

Without all hope of day!" --Samson Agonistes, Lines 80-2.

"It was that fatal and perfidious bark, Built in th' eclipse, and rigg'd with curses dark, That sunk so low that sacred heart of thine." --Lycidas, Lines 100-2.

Pope, in the following lines, may be presumed to mean that the covering up of the Sun by the Moon, during a total eclipse, results in the Moon becoming visible, at the cost of the Sun's disappearance:--

"For Envy'd wit, like Sol eclips'd, makes known Th' opposing body's grossness, not its own." --Essay on Criticism, Lines 469-70.

I have not attempted to pursue this matter through the pages of our modern poets, but it is not unlikely that Scott and Tennyson (especially) would have something on the subject of eclipses.

FOOTNOTES:

[Footnote 166: Paradise Lost, Book i., lines 594-9.]

[Footnote 167: The Astronomy of Milton, p. 259.]

CHAPTER XX.

BRIEF HINTS TO OBSERVERS OF ECLIPSES OF THE SUN.

A few words (they must be few for lack of space) may usefully be added, by way of advice, to persons proposing to choose a suitable locality at which to station themselves for viewing a total eclipse of the Sun. To begin with, of course they ought to get as close as possible to the central line, say within 10 or 20 miles at the most; this matter settled, the next important point is to find out where the duration of the totality will be longest, coupled with the Sun at its maximum elevation above the horizon (to escape the influence of mists and fogs). No advice, properly so-called, can be given on these points, because they depend on the special circumstances of every eclipse, and must be ascertained ad hoc from the Nautical Almanac.

In anticipation of a forthcoming eclipse, it is very important to know beforehand the probabilities of weather. If the locus in quo of an expected eclipse is in a civilised country, there will generally not be much difficulty in obtaining a certain amount of information as to this 6 or 12 months in advance. But inasmuch as total eclipses of the Sun, and often the best of them, are visible only in uncivilised countries or over trackless wastes, the problem becomes a complicated and anxious one. In such cases it is exceedingly desirable, where competent observers (including money) are available, that preliminary notes of weather should be made for a year or even two years in advance. There is in one sense no difficulty as to this, for all the mathematical local elements of every eclipse are always made public three or four years in advance through the pages of books like the Nautical Almanac, the Connaissance des Temps, the Berliner Jahrbuch, &c. One difficulty always confronts every eclipse expedition. If an out-of-the-way part of the world has to be visited, accessible by sea, transport from England, say, to the foreign shore is not usually a matter of difficulty, because Government ships are often placed at the disposal of astronomers. But the gravest difficulties often have to be faced after the arrival at the foreign shore, and for this reason. Every sea coast is, as a general rule applicable to the whole world, bad for astronomical observations. The problem then which has to be solved is, how best to get away from the coast inland to a high hill, and to find the means of transporting thither heavy packing-cases of instruments, personal luggage, creature comforts, and, if needs be, tents and the other accessories of camp life. Let not the reader of either sex take fright at the idea of sleeping under a tent. I speak with considerable experience when I say that, given fine or fairly fine weather, nothing is more enjoyable in a temperate climate. Under the term "creature comforts" I mean such things as tinned soups and preserved provisions which nowadays can so easily be purchased everywhere in England, and of such good quality. I would recommend these being taken even when the eclipse traveller expects to be lodged in the dwelling-places of civilised nations. Of course, if in order to see his eclipse he has to go into the wilds of America, Asia, or Africa, he must start fully equipped with all those personal impedimenta which will be found scheduled in the books mentioned in the footnote.[168]

FOOTNOTES:

[Footnote 168: The Tourists' Pocket-Book, 1s. (Philip); F. Galton's Art of

Travel, 7s. 6d. (Murray); Royal Geographical Society's Hints to Travellers, 5s.

CHAPTER XXI.

TRANSITS AND OCCULTATIONS.

No book professing to deal with eclipses would be complete without a few words of mention of "transits" and "occultations." A transit is the passing of a primary planet across the Sun, or of a secondary planet (i.e. satellite) across its primary, whilst an occultation is the concealment of a star by the Moon, or of a secondary planet (i.e. satellite) by its primary. A little thought given to this definition will make it clear that a transit is essentially the same in principle as an eclipse of the Sun by the Moon--one body comes in front of another, and the former conceals in succession parts of the latter.

Practically the word "transit" in this connection is more especially applied to passages of the inferior planets, Mercury and Venus, across the Sun, or of the satellites of Jupiter across the disc of Jupiter, whilst the word "occultation" more particularly calls to mind the concealment of a star (apparently a little body) by the Moon (apparently a big body) or of a satellite of Jupiter (a little body) by Jupiter (a big body), the star and the satellite in each respective case passing behind the occulting body and being concealed for a shorter or longer time. Commonly the occulted body will remain hidden for an hour or two, more or less. In the case of Jupiter the satellites of that planet may also, on occasions, be seen to undergo eclipse in the shadow cast by Jupiter itself. An eclipse of a Jovian satellite is therefore on all fours in principle the same as an eclipse of the Moon, caused, as we know, by the Moon passing for a short time into the dark shadow cast by the Earth. The conditions just laid down in respect of Jupiter and its satellites also find a counterpart in the case of the satellites of Saturn, but whilst these phenomena are incessantly occurring and visible in the case of Jupiter, they are exceedingly rare in the case of Saturn owing to its greater distance and the difficulty of seeing most of its satellites because of their small apparent size.

Having regard to the circumstance that transits of Mercury and Venus only happen at intervals of many years, it is not worth while for the purposes of this work to devote any great amount of space to them. In point of fact, whilst the next three transits of Mercury are as remote as 1907, 1914 and

1924, there will be no transit of Venus at all during the 20th century; not another indeed until A.D. 2004.

From the standpoint of an amateur astronomer the various phenomena which attend the movements of the satellites of Jupiter, constitute an endless variety of interesting scenes, which are the more deserving of attention in that they can be followed with the aid of a telescope of very moderate size and capabilities.[169]

Occultations of planets and stars by the Moon may also be recommended to the notice of the owners of small telescopes as events which are constantly happening and which may be readily observed. The Moon being rapidly in motion it will happen in point of fact that stars are occulted by it, one may say every day, but of course the Moon's light entirely blots out the smaller stars and only those as large as, say, about the 5th magnitude are as a rule worth trying to see in this connection. A table of the occultations of such stars, copied from the Nautical Almanac, will be found in such almanacs as Whitaker's and the British. If such a table is consulted it will be found that never does a lunation pass without a few stars being noted as undergoing occultation, and now and then a planet. An occultation of a planet is obviously still more interesting than that of a star.

From the epoch of New to Full Moon the Moon moves with its dark edge foremost from the epoch of Full to New with its illuminated edge foremost. During therefore the first half of a lunation the objects occulted disappear at the dark edge and reappear at the illuminated edge, during the second half of a lunation things are vice vers? The most interesting time for watching occultations is with a young Moon no more than, say, from 2 to 6 days old, because under such circumstances the star occulted is suddenly extinguished at a point in the sky where there seems nothing to interfere with it.

FOOTNOTES:

[Footnote 169: For details as to these matters, see my Handbook of Astronomy, 4th ed., vol. i. pp. 186-196.]

www.ingramcontent.com/pod-product-compliance
Lightning Source LLC
Chambersburg PA
CBHW070859180526
45168CB00005B/1875